BORDER SECURITY: BARRIERS ALONG THE U.S. INTERNATIONAL BORDER

BORDER SECURITY: BARRIERS ALONG THE U.S. INTERNATIONAL BORDER

BLAS NUÑEZ-NETO
AND
MICHAEL JOHN GARCIA

Nova Science Publishers, Inc.
New York

LIBRARY OF CONGRESS CATALOGING-IN-PUBLICATION DATA

Available Upon Request
ISBN 978-1-60692-171-5

Published by Nova Science Publishers, Inc. ✦New York

Contents

PREFACE

Congress has repeatedly shown interest in examining and expanding the barriers being deployed along the U.S. international land border. The 109th Congress passed a number of laws affecting these barriers, and oversight of these laws and of the construction process may be of interest to the 110th Congress. The United States Border Patrol (USBP) deploys fencing, which aims to impede the illegal entry of individuals, and vehicle barriers, which aim to impede the illegal entry of vehicles (but not individuals) along the border. The USBP first began erecting barriers in 1990 to deter illegal entries and drug smuggling in its San Diego sector. The ensuing 14 mile-long San Diego "primary fence" formed part of the USBP's "Prevention Through Deterrence" strategy, which called for reducing unauthorized migration by placing agents and resources directly on the border along population centers in order to deter would-be migrants from entering the country. In 1996, Congress passed the Illegal Immigration Reform and Immigrant Responsibility Act which, among other things, explicitly gave the Attorney General (now the Secretary of the Department of Homeland Security) broad authority to construct barriers along the border and authorized the construction of a secondary layer of fencing to buttress the completed 14 mile primary fence. Construction of the secondary fence stalled due to environmental concerns raised by the California Coastal Commission. In 2005, Congress passed the REAL ID Act that authorized the Secretary of the Department of Homeland Security (DHS) to waive all legal requirements in order to expedite the construction of border barriers. DHS has announced it will use this waiver authority to complete the San Diego fence. The Secure Fence Act of 2006 directed DHS to construct 850 miles of additional border fencing. This requirement was subsequently modified by the Consolidated Appropriations Act, 2008 (P.L. 110-161), which was enacted into law on December 26, 2007. The act requires the Secretary of Homeland Security to construct fencing along not less than 700 miles of the southwest border.

While the San Diego fence, combined with an increase in agents and other resources in the USBP's San Diego sector, has proven effective in reducing the number of apprehensions made in that sector, there is considerable evidence that the flow of illegal immigration has adapted to this enforcement posture and has shifted to the more remote areas of the Arizona desert. Nationally, the USBP made 1.2 million apprehensions in 1992 and again in 2004, suggesting that the increased enforcement in San Diego sector has had little impact on overall apprehensions. In addition to border fencing, the USBP deploys both permanent and temporary vehicle barriers to the border. Temporary vehicle barriers are typically chained together and can be moved to different locations at the USBP's discretion. Permanent vehicle barriers are embedded in the ground and are meant to remain in one location.

A number of policy issues concerning border barriers generally and fencing specifically may be of interest to Congress, including, but not limited, to their effectiveness, costs versus benefits, location, design, environmental impact, potential diplomatic ramifications, and the costs of acquiring the land needed for construction.

Border Security: Barriers Along the U.S. International Border

Background

Within the Department of Homeland Security's (DHS) Customs and Border Protection (CBP), the U.S. Border Patrol (USBP) is charged with securing our nation's land and maritime borders between official ports of entry (POE) to deter and interdict terrorists, weapons of mass destruction, and aliens attempting to enter the country unlawfully. In order to discharge its duties, the USBP deploys personnel, technology, and tactical infrastructure such as vehicle barriers and fencing. Fencing is erected on the border to impede the illegal entry of unauthorized aliens, while vehicle barriers are designed to impede the entry of vehicles but do not impede the entry of individuals. This report will analyze the barriers that are currently being constructed and maintained along the border by the USBP, including historical and future cost estimates and the policy issues involved. Because the current debate has largely focused on the deployment of fencing to the border, this report will focus on the policy issues surrounding the construction of border fencing. However, information concerning the kinds of vehicle barriers being deployed at the border will be provided where available.

Using the broad powers granted to the Attorney General (AG) to control and guard the U.S. border,[1] the USBP began erecting a barrier known as the "primary fence" directly on the border in 1990 to deter illegal entries and drug smuggling in its San Diego sector.[2] The San Diego fence formed part of the USBP's "Prevention Through Deterrence" strategy,[3] which called for reducing unauthorized migration by placing agents and resources directly on the border along population centers in order to deter would-be migrants from entering the country. The San Diego primary fence was completed in 1993, covering the first 14 miles of the border from the Pacific Ocean. The fence was constructed of 10-foot-high welded steel army surplus landing

---

[*] Excerpted from CRS Report # RL33659; January 8, 2008

[1] 8 U.S.C. §1103 (a)(5). Although the law still cites to the Attorney General, the authorities granted by this section now appear to rest with the Secretary of DHS. See The Homeland Security Act of 2002, P.L. 104-208, §§102(a), 441, 1512(d) and 1517 (references to the Attorney General or Commissioner in statute and regulations are deemed to refer to the Secretary of DHS).

[2] For more information on the San Diego border fence, please refer to CRS Report RS22026, *Border Security: The San Diego Fence*, by Blas Nuñez-Neto and Michael John Garcia.

[3] For an expanded discussion of the USBP, please refer to CRS Report RL32562, *Border Security: The Role of the U.S. Border Patrol*, by Blas Nuñez-Neto.

mats[4] with the assistance of the Corps of Engineers and the California National Guard. In addition to the 14 miles of primary fencing erected in its San Diego sector, the USBP maintains stretches of primary fencing in several other sectors along the southwest border, including Yuma, Tucson, El Centro, and El Paso.

In 1996, Congress passed the Illegal Immigration Reform and Immigrant Responsibility Act (IIRIRA), which, among other things, explicitly gave the Attorney General broad authority to construct barriers along the border and authorized the Immigration and Naturalization Service (INS) to construct a secondary layer of fencing to buttress the completed 14 mile primary fence.[5] Construction of the secondary fence stalled after 9.5 miles had been completed due to environmental concerns raised by the California Coastal Commission (CCC). In 2005, Congress passed the REAL ID Act, which, among other things, authorized the Secretary of the Department of Homeland Security (DHS) to waive all legal requirements to expedite the construction of border barriers.[6] In 2006, Congress passed the Secure Fence Act, which, among other things, directed DHS to construct five separate stretches of fencing along the southern border totaling 850 miles.[7] This requirement was modified by provisions in Division E of H.R. 2764, the Consolidated Appropriations Act, 2008 (P.L. 110-161), which was enacted into law on December 26, 2007. The Secretary of Homeland Security is now required to construct reinforced fencing along not less than 700 miles of the southwest border, in locations where fencing is deemed most practical and effective.

In addition to border fencing, the USBP deploys both permanent and temporary vehicle barriers at the border. Vehicle barriers are meant to stop the entry of vehicles, but not people, into the United States. Temporary vehicle barriers are typically chained together and can be moved to different locations at the USBP's discretion. Permanent vehicle barriers are embedded in the ground and are meant to remain in one location. The USBP is currently erecting a 150 mile stretch of vehicle barriers in conjunction with the National Park Service near Yuma, Arizona.

The San Diego Border Primary Fence

The USBP's San Diego sector extends along the first 66 miles from the Pacific Ocean of the international border with Mexico, and covers approximately 7,000 square miles of territory. Located north of Tijuana and Tecate, Mexican cities with a combined population of more than two million people, the sector features no

---

[4] U.S. Government Accountability Office, *Border Control — Revised Strategy Is Showing Some Positive Results*, GAO/GGD-95-30, January 31, 1995. (Hereafter referred to as GAO Report 95-30.)

[5] See P.L. 104-208, Div. C. IIRIRA was passed as part of the Omnibus Consolidated Appropriations Act of 1997.

[6] P.L. 109-13.

[7] From CBP Congressional Affairs, September 25, 2006.

natural barriers to entry by unauthorized migrants and smugglers.[8] As a result of this geographical reality and in response to the large numbers of unauthorized aliens crossing the border in the area, in 1990 the USBP began erecting a physical barrier to deter illegal entries and drug smuggling. The ensuing "primary" fence covered the first 14 miles of the border, starting from the Pacific Ocean, and was constructed of 10-foot-high welded steel.[9]

Operation Gatekeeper

The primary fence, by itself, did not have a discernible impact on the influx of unauthorized aliens coming across the border in San Diego. As a result of this, Operation Gatekeeper was officially announced in the San Diego sector on October 1, 1994. The chief elements of the operation were large increases in the overall manpower of the sector, and the deployment of USBP personnel directly along the border to deter illegal entry. The strategic plan called for three tiers of agent deployment. The first tier of agents was deployed to fixed positions on the border. The agents in this first tier were charged with preventing illegal entry, apprehending those who attempted to enter, and generally observing the border. A second tier of agents was deployed north of the border in the corridors that were heavily used by illegal aliens. The second tier of agents had more freedom of movement than the first tier and were charged with containing and apprehending those aliens who made it past the first tier. The third tier of agents were typically assigned to man vehicle checkpoints further inland to apprehend the traffic that eluded the first two tiers. As the Department of Justice Inspector General report notes, "given Gatekeeper's deterrence emphasis, many agents were assigned to first-tier, fixed positions along the border. These agents were instructed to remain in their assigned positions rather than chase alien traffic passing through adjacent areas. Prior to Gatekeeper, such stationary positions were relatively rare."[10]

Operation Gatekeeper resulted in significant increases in the manpower and other resources deployed to San Diego sector. Agents received additional night vision goggles, portable radios, and four-wheel drive vehicles, and light towers and seismic sensors were deployed.[11] According to the former INS, between October 1994 and June of 1998, San Diego sector saw the following increases in resources:

- USBP agent manpower increased by 150%;
- Seismic sensors deployed increased by 171%;
- Vehicle fleet increased by 152%.
- Infrared night-vision goggles increased from 12 to 49;

---

[8] U.S. Department of Justice, Office of the Inspector General, *Operation Gatekeeper: An Investigation Into Allegations of Fraud and Misconduct*, July 1998, available at [http://www.usdoj.gov/oig/special/9807/gkp01.htm#P160_18689].

[9] GAO Report 95-30.

[10] U.S. Department of Justice, Office of the Inspector General, *Operation Gatekeeper: An Investigation Into Allegations of Fraud and Misconduct*, July 1998, at [http://www.usdoj.gov/oig/special/9807/index.htm]. (Hereafter referred to as DOJ-OIG Gatekeeper Report.)

[11] DOJ-OIG Gatekeeper Report.

- Permanent lighting increased from 1 mile to 6 miles, and 100 portable lighting platforms were deployed;
- Helicopter fleet increased from 6 to 10.[12]

As a result of the increase in resources and the new strategy that were the main components of Operation Gatekeeper, the USBP estimated in 1998 that the entire 66 miles of border patrolled by the San Diego sector's agents could be brought under control in five years.[13]

Sandia National Laboratory Study

According to CBP, the primary fence, in combination with various USBP enforcement initiatives along the San Diego border region (i.e., Operation Gatekeeper), proved to be successful but fiscally and environmentally costly.[14] For example, as unauthorized aliens and smugglers breached the primary fence and attempted to evade detection, USBP agents were often forced to pursue the suspects through environmentally sensitive areas. It soon became apparent to immigration officials and lawmakers that the USBP needed, among other things, a "rigid" enforcement system that could integrate infrastructure (i.e., a multi-tiered fence and roads), manpower, and new technologies to further control the border region.

The concept of a three-tiered fence system was first recommended by a 1993 Sandia Laboratory study commissioned by the former Immigration and Naturalization Service (INS). According to the Sandia study, the use of multiple barriers in urban areas would increase the USBP's ability to discourage a significant number of illegal border crossers, to detect intruders early and delay them as long as possible, and to channel a reduced number of illegal border crossers to geographic locations where the USBP was better prepared to deal with them.[15] The Sandia study further noted that segments of the border could not be controlled at the immediate border due to the ruggedness of the terrain, and recommended the use of highway checkpoints in those areas to contain aliens after they had entered the country illegally.[16] The study concluded that aliens attempting to enter the United States from Mexico had shown remarkable resiliency in bypassing or destroying obstacles in their path, including the existing primary fence, and postulated that "[a] three-fence barrier

---

[12] U.S. Department of Justice, Immigration and Naturalization Service, "Operation Gatekeeper Fact Sheet," July 14, 1998.

[13] DOJ-OIG Gatekeeper Report.

[14] See California Coastal Commission, *W 13a Staff Report and Recommendation on Consistency Determination*, CD-063-03, October 2003 [hereinafter "CCC *Staff Report*"], at 14-16 (stating that construction of the primary fence significantly assisted the USBP's efforts in deterring smuggling attempts via drive-throughs using automobiles and motorcycles).

[15] GAO 95-30, p. 13.

[16] GAO 95-30, p. 13.

system with vehicle patrol roads between the fences and lights will provide the necessary discouragement."[17]

Congressional Border Barrier Legislation

As previously mentioned, the INS constructed the primary fencing in San Diego using the broad authority granted to the AG in order to guard and control the U.S. border by the Immigration and Nationality Act (INA).[18] In 1996, Congress expressly authorized the AG to construct barriers at the border for the first time in the Illegal Immigration Reform and Immigrant Responsibility Act (IIRIRA).[19] This legislation has subsequently been amended on several occassions.

Section 102 of IIRIRA — Improvement of Barriers at the Border

Section 102 of IIRIRA concerned the improvement and construction of barriers at our international borders. As originally enacted, § 102(a) appeared to give the AG[20] broad authority to install additional physical barriers and roads "in the vicinity of the United States border to deter illegal crossings in areas of high illegal entry into the United States." The phrase "vicinity of the United States border" was not defined in the INA or in immigration regulations. The section also did not stipulate what specific characteristics would designate an area as one of "high illegal entry."

As originally enacted, § 102(b) mandated that the AG construct a barrier in the border area near San Diego. Specifically, §102(b) directed the AG to construct a three-tiered barrier along the 14 miles of the international land border of the U.S., starting at the Pacific Ocean and extending eastward. Section 102(b) ensured that the AG will build a barrier, pursuant to his broader authority in §102(a), near the San Diego area, although there is some debate concerning whether IIRIRA required *continuous* triple fencing and roads for the entire 14-mile corridor.[21] IIRIRA § 102(b) also provided authority for the acquisition of necessary easements, required certain safety features be incorporated into the design of the fence, and authorized a

---

[17] Peter Andreas, "The Escalation of U.S. Immigration Control in the Post-NAFTA Era," *Political Science Quarterly*, vol. 113, no. 4, winter 1998-1999, p. 595.

[18] 8 U.S.C. §1103 (a)(5).

[19] P.L. 104-208, §102.

[20] The Consolidated Appropriations Act, 2008 (P.L. 110-161) amended IIRIRA § 102 to expressly refer to the Secretary of Homeland Security, rather than the Attorney General. Although IIRIRA § 102 previously referred to the Attorney General, the authorities granted by this section nonetheless appeared to rest with the Secretary of DHS following the enactment of the Homeland Security Act of 2002. See P.L. 104-208, §§102(a), 441, 1512(d) and 1517 (references to the Attorney General or Commissioner in statute and regulations are deemed to refer to the Secretary).

[21] See CCC, *Staff Report*, *supra* note 14, at pp. 7 nt. 2 and 23 nt. 4.

total appropriation not to exceed $12 million to carry out the section.[22] The Secure Fence Act of 2006 (P.L. 109-367) amended IIRIRA § 102(b) by *removing* the specific provisions authorizing construction of the San Diego fence (though not the provisions concerning fence safety features, easements, or appropriations) and adding provisions authorizing five stretches of two-layered reinforced fencing, totaling roughly 850 miles, along the southwest border.[23] IIRIRA § 102(b) was again amended by the Consolidated Appropriations Act, 2008 (P.L. 110-161). The Secretary of Homeland Security is now required to construct reinforced fencing along not less than 700 miles of the southwest border, in locations where fencing is deemed most practical and effective.[24] The Consolidated Appropriations Act also amended IIRIRA § 102(b) to authorize the appropriation of "sums as may be necessary to carry out this subsection." Although IIRIRA § 102(b) no longer contains a specific authorization for the San Diego fence, the project appears permissible under the general fence authorization contained in IIRIRA §102(a).

As originally enacted, IIRIRA § 102(c) waived the Endangered Species Act (ESA) of 1973 (16 U.S.C. §§1531 *et seq.*) and the National Environmental Policy Act (NEPA) of 1969 (42 U.S.C. §§4321 *et seq.*), to the extent the AG determined necessary, in order to ensure expeditious construction of the barriers authorized to be constructed under §102. The waiver authority in this provision appeared to apply both to barriers that may be constructed *in the vicinity of the border* and to the barrier that was to be constructed near the San Diego area. The INS (and CBP after 2003) never exercized this original waiver authority, instead choosing to comply with the NEPA and the ESA. The INS published a Final Environmental Impact Study pursuant to NEPA and received a non-jeopardy Biological Opinion from the U.S. Fish and Wildlife Service under the ESA.[25] This waiver authority was expanded in the 109th Congress by the REAL ID Act, which will be discussed in greater detail subsequently, and DHS has exercised this expanded waiver authority in order to continue construction of the San Diego border fence, as well as physical barriers and roads along the southwest border.

Section 102(d) also provided the AG with various land acquisition authorities. In 2002, Congress authorized the AG to use INS funds to purchase land for enforcement fences and to construct the fences.[26]

---

[22] The actual costs associated with constructing the San Diego fence have been considerably greater than anticipated by IIRIRA and will be discussed in more detail later in this report.

[23] For more detailed discussion of the Secure Fence Act, see *infra* at 29.

[24] For more detailed discussion of the amendments made by the Consolidated Appropriations Act, see *infra* at 26-27.

[25] Department of Homeland Security, *Environmental Impact Statement for the Completion of the 14-mile Border Infrastructure System, San Diego, California* (July 2003) [hereinafter "EIS, San Diego Border Fence"].

[26] P.L.107-273, §201(a).

Expansion of Waiver Authority under the REAL ID Act

As mentioned above, pursuant to the REAL ID Act of 2005 (P.L. 109-13, Division B),[27] the Secretary of DHS was given broad authority to waive legal requirements that might otherwise delay the construction of the security barriers described under § 102 of IIRIRA. Specifically, the Secretary of DHS is authorized to waive *all legal requirements* necessary to ensure expeditious construction of these security barriers.[28] Such waivers are effective upon publication in the *Federal Register*. Federal district courts are provided with exclusive jurisdiction to review claims alleging that the actions or decisions of the Secretary violate the U.S. Constitution, and district court rulings may be reviewed only by the Supreme Court.

The scope of this waiver authority is substantial. Whereas IIRIRA had previously authorized the waiver of NEPA and ESA requirements, the REAL ID Act authorizes the waiver of *all* legal requirements determined necessary by the Secretary for the expeditious construction of authorized barriers, and only allows judicial review for constitutional claims. This waiver authority appears to apply to *all* barriers that may be constructed under IIRIRA — that is, both to barriers constructed in the vicinity of the border in areas of high illegal entry and to the barrier that is to be constructed near the San Diego area. Furthermore, these claims can only be appealed to the Supreme Court (i.e., there is no intermediate appellate review), whose review is discretionary.

Some have expressed concern with the apparent breadth of the waiver provision and the limited scope of judicial review of waiver decisions. As passed into law, the REAL ID Act waiver provision begins with the arguably ambiguous "notwithstanding any other law" phrase[29] and allows the waiver of all "legal

[27] For more information on the REAL ID Act, see CRS Report RL32754, *Immigration: Analysis of the Major Provisions of the REAL ID Act of 2005*, by Michael John Garcia, Margaret Mikyung Lee, and Todd Tatelman.

[28] As initially introduced as H.R. 418, the REAL ID Act *required* the Secretary of DHS to waive *all laws* necessary to ensure expeditious construction of the security barriers. H.R. 418 was passed by the House as a stand-alone piece of legislation, but was subsequently attached as an amendment to House-passed H.R. 1268, the emergency supplemental appropriations bill for FY2005. During conference, language was revised in H.R. 1268, so that the Secretary was authorized, but not required, to waive all "legal requirements" (instead of "all laws") deemed necessary to ensure construction of the security barriers. The conferees also added provisions to the REAL ID Act which made waiver decisions effective upon publication in the *Federal Register*, and permitted federal court review of waiver decisions only in limited circumstances. The conference version of H.R. 1268 was enacted on May 11, 2005.

[29] Some courts, for instance, have found the "notwithstanding" phrase not dispositive in determining the preemptive effect of a statute. See, e.g., *E.P. Paup v. Director*, OWCP, 999 F.2d 1341, 1348 (9th Cir. 1993); *Oregon Natural Resources Council v. Thomas*, 92 F.3d 792, 796 (9th Cir. 1996). But see *Puerto Rico v. M/V Emily S.*, 132 F.3d 818 (1st Cir. 1997); *Schneider v. United States*, 27 F.3d 1327 (8th Cir. 1994).

requirements." Although the term "legal requirement" is not defined, it cannot grant the Secretary the authority to unilaterally waive a person's constitutional rights.[30]

The provision has been construed by Secretary Chertoff to the waiver of laws in their entirety, along with regulations and requirements deriving from or relating to such laws. Congress commonly waives preexisting laws, but the new waiver provision uses language and a combination of terms not typically seen in law. Most waiver provisions have contained qualifying language that (1) exempts an action from other requirements contained in the act that authorizes the action, (2) specifically delineates the laws to be waived, or (3) waives a grouping of similar laws. Also common are waiver provisions that contain reporting requirements or restrictions which appear to limit their breadth.[31] One waiver authority that appears analogous to that contained in the REAL ID Act is § 203 of the Trans-Alaska Pipeline Authorization Act, as amended, which authorizes the Secretary of the Interior to waive all procedural requirements in law related to the construction of the Trans-Alaska pipeline and limits judicial review to constitutional claims.[32]

Although some argue that the waiver authority can extend to any law, including those seemingly unrelated to building a fence (e.g., civil rights or child labor laws), the provision is tempered by the requirement that the Secretary must determine the law (subject to the waiver) is necessary "to ensure expeditious construction" of the barriers. In other words, the Secretary may be confined to laws that, in effect, will impede the construction of the fence — not those that only tangentially relate to or do not necessarily interfere with construction. For example, because child labor laws would not prevent the Secretary from expeditiously constructing the fence, it follows that the Secretary does not have the authority to waive these protections. This interpretation is buttressed by the legislative history of the REAL ID Act, which indicates that several Members called for the waiver provision because of laws that were complicating and ultimately preventing the completion of the fence.[33] The decision to waive a law, nonetheless, is solely in the Secretary's discretion. Until such time that DHS waives an applicable law, however, it must follow all legal requirements normally imposed on federal agencies.

On September 22, 2005, a notice was issued in the *Federal Register* indicating that Secretary Chertoff, acting pursuant to the authority provided under the REAL ID

---

[30] "[T]he Constitution is filled with provisions that grant Congress or the States specific power to legislate in certain areas," Justice Black wrote for the Court, but "these granted powers are always subject to the limitations that they may not be exercised in a way that violates other specific provisions of the Constitution." Williams v. Rhodes, 393 U.S. 23, 29 (1968).

[31] Some of these waiver provisions grant the President or the head of an Executive agency the authority to waive a law[s] if deemed necessary in the *national interest* or in the interest of *national defense*. See, e.g., 10 U.S.C. §1107(a); 22 U.S.C. §2375(d); 29 U.S.C. §793; 42 U.S.C. §6212(b); 42 U.S.C. §6393(a)(2); 50 U.S.C. §2426(e). Examples of waiver authority with a congressional notification element include 15 U.S.C. §719f; 22 U.S.C. §2378; 22 U.S.C. §2371; and 41 U.S.C. §413.

[32] P.L. 93-153, Title II, § 203 (1973); 43 U.S.C. §1652(c)-(d).

[33] 151 Cong. Rec. H557 (daily ed. February 10, 2005).

Act, had exercised waiver authority over various legal requirements in order to ensure the expeditious construction of the San Diego border fence.[34] A listing of laws waived by the Secretary can be found in **Appendix VI**. A notice was also published on January 19, 2007, indicating that the Secretary was waiving various legal requirements in order to ensure the expeditious construction of physical barriers and roads in the vicinity of U.S. border area known as the Barry M. Goldwater Range (BMGR), in southwestern Arizona.[35] A listing of the federal laws waived by the Secretary pursuant to this notice can be found in **Appendix VII**.

On October 5, 2007, Defenders of Wildlife and the Sierra Club brought suit in the U.S. District Court for the District of Columbia seeking a temporary restraining order enjoining DHS from border fence and road-building activities in the San Pedro Riparian National Conservation Area, located in the vicinity of the U.S. border in southeastern Arizona.[36] On October 10, 2007, the presiding district court judge issued a temporary restraining order (TRO) halting fence construction activities in the Conservation Area, finding the relevant federal agencies had failed to carry out an environmental assessment as legally required. On October 26, 2007, a notice was published in the *Federal Register* indicating that the Secretary of Homeland Security had exercised waiver authority over various legal requirements in order to ensure the expeditious construction of physical barriers or roads through the San Pedro Riparian National Conservation Area (including any and all lands covered by the TRO),[37] thereby enabling the DHS to resume fence construction. A listing of the federal laws waived by the Secretary pursuant to this notice can be found in **Appendix VIII**. Defenders of Wildlife and the Sierra Club subsequently filed an amended complaint on November 1, 2007, challenging the constitutionality of DHS's waiver authority.[38]

---

[34] Dept. of Homeland Security, "Determination Pursuant to Section 102 of the Illegal Immigration Reform and Immigrant Responsibility Act of 1996 as Amended by Section 102 of the REAL ID Act of 2005," 70 *Federal Register* 55622-02, September 22, 2005 [hereinafter "DHS Notice"].

[35] Dept. of Homeland Security, "Determination Pursuant to Section 102 of the Illegal Immigration Reform and Immigrant Responsibility Act of 1996 as Amended by Section 102 of the REAL ID Act of 2005 and as Amended by the Secure Fence Act of 2006," 72 *Federal Register* 2535-01, January 19, 2007.

[36] Defenders of Wildlife v. Bureau of Land Management, Case 1:07-cv-01801-ESH (D.D.C. 2007). Plaintiffs' request for a temporary restraining order can be viewed at [http://www.defenders.org/resources/publications/programs_and_policy/in_the_courts/san_pedro_border_wall_tro_filing.pdf].

[37] Dept. of Homeland Security, "Determination Pursuant to Section 102 of the Illegal Immigration Reform and Immigrant Responsibility Act of 1996 as Amended by Section 102 of the REAL ID Act of 2005 and as Amended by the Secure Fence Act of 2006," 72 *Federal Register* 60870-01,October 26, 2007.

[38] Plaintiffs' amended complaint can be viewed at [http://www.defenders.org/resources/publications/programs_and_policy/in_the_courts/san_pedro_border_wall_amended_complaint.pdf].

On December 18, 2007, the district court issued an opinion rejecting plaintiffs' constitutional challenge and granting DHS's motion to dismiss the case.[39]

The San Diego Sandia Fence

In 1996, construction began on the secondary fence that had been recommended by the Sandia study with congressional approval. The new fence was to parallel the fourteen miles of primary fence already constructed on land patrolled by the Imperial Beach Station of the San Diego sector, and included permanent lighting as well as an access road in between the two layers of fencing. Of the 14 miles of fencing authorized to be constructed by IIRIRA, nine miles of the triple fence had been completed by the end of FY2005. Two sections, including the final three mile stretch of fence that leads to the Pacific Ocean, have not been finished.

The California Coastal Commission

In order to finish the fence, the USBP proposed to fill a deep canyon known as "Smuggler's Gulch" with over two million cubic yards of dirt. The triple-fence would then be extended across the filled gulch. California's Coastal Commission (CCC), however, objected to and essentially halted the completion of the fence in February 2004, because it determined that CBP had not demonstrated, among other things, that the project was consistent "to the maximum extent practicable" with the policies of the California Coastal Management Program — a state program approved under the federal Coastal Zone Management Act (CZMA) (16 U.S.C. §§1451-1464).[40] The CZMA requires federal agency activity within or outside the coastal zone that affects any land or water use or natural resource of the coastal zone to be carried out in a manner that is consistent to the maximum extent practicable with the policies of an approved state management program.[41] If a federal court finds a federal activity to be inconsistent with an approved state program and the Secretary of DHS (Secretary) determines that compliance is unlikely to be achieved through mediation, the President may exempt from compliance the activity if the President determines that the activity is in the "paramount interest of the United States."[42]

According to the CCC, CBP did not believe that it could make further environmental concessions and still comply with IIRIRA. The CCC held that Congress did not specify a particular design in the IIRIRA, and that CBP failed to present a convincing argument that the less environmentally damaging alternative projects it rejected would have prevented compliance with the IIRIRA. Specifically,

---

[39] Defenders of Wildlife v. Chertoff, Case 1:07-cv-01801-ESH, 2007 U.S. Dist. LEXIS 92648 (D.D.C. December 18, 2007)

[40] See CCC, *Staff Report*, at 5-7. After California's Coastal Management Plan was approved by the National Oceanic and Atmospheric Administration pursuant to the CZMA in 1977, apparently all federal activities affecting coastal zone resources in California became subject to the CCC's regulatory purview.

[41] 16 U.S.C. §1456(c).

[42] 16 U.S.C. §1456(c)(1)(B).

the CCC was concerned with the potential for significant adverse effects on (1) the Tijuana River National Estuarine Research and Reserve; (2) state and federally listed threatened and endangered species; (3) lands set aside for protection within California's Multiple Species Conservation Program; and, (4) other aspects of the environment. In response to the CCC's findings, Congress expanded the waiver authority in the REAL ID Act, described in more detail below, in order to allow DHS to waive the CZMA, among other things.

Current Status of the San Diego Triple Fence

As previously discussed, DHS announced in September 2005 that it was applying its waiver authority established by the REAL ID Act to facilitate the completion of the San Diego fence.[43] The military has now begun the process of upgrading and rebuilding the San Diego border fence. Congress appropriated $31 million in FY2007 for construction of the remaining 4.5 miles of the San Diego fence.[44] DHS has begun construction on the final 4.5 miles of the San Diego fence, beginning the process of filling in the area known as Smuggler's Gulch.[45]

The San Diego Fence and USBP Apprehensions

Apprehension statistics have long been used as a performance measure by the USBP. However, the number of apprehensions may be a misleading statistic for several reasons, including the data's focus on events rather than people[46] and the fact that there are no reliable estimates for how many aliens successfully evade capture. This makes it difficult to establish a firm correlation between the number of apprehensions in a given sector and the number of people attempting to enter through that sector. While caution should be taken when attempting to draw conclusions about the efficacy of policy initiatives based solely on apprehensions statistics, *they remain the most reliable way to codify* trends in illegal migration along the border.

The San Diego fence spans two border patrol stations within the San Diego sector: Imperial Beach station and Chula Vista station. As previously noted, the primary fence was constructed in those two stations beginning in FY1990; the secondary fence was constructed beginning in FY1996. **Figure 1** shows the stark decrease in apprehensions at the Imperial Beach station *from* FY1992 *to* FY2004. The majority of the decrease occurred in the four year period from FY1995 through FY1998 and coincided with Operation Gatekeeper, which as previously noted combined the construction of fencing along the border with an increase in agents and other resources deployed directly along the border. For the period from FY1998 to

---

[43] DHS Notice, *supra* note 35.

[44] H.Rept. 109-699, p. 130.

[45] Interview with CBP Congressional Affairs, January 26 , 2006.

[46] If the same person is apprehended multiple times attempting to enter the country in one year, each apprehension will be counted separately by the USBP in generating their apprehension statistics. This means that apprehension statistics may overstate the number of aliens apprehended each year.

FY2004, apprehensions at the Imperial Beach station averaged about 14,000 each year.

Figure 1. Imperial Beach Station Apprehensions

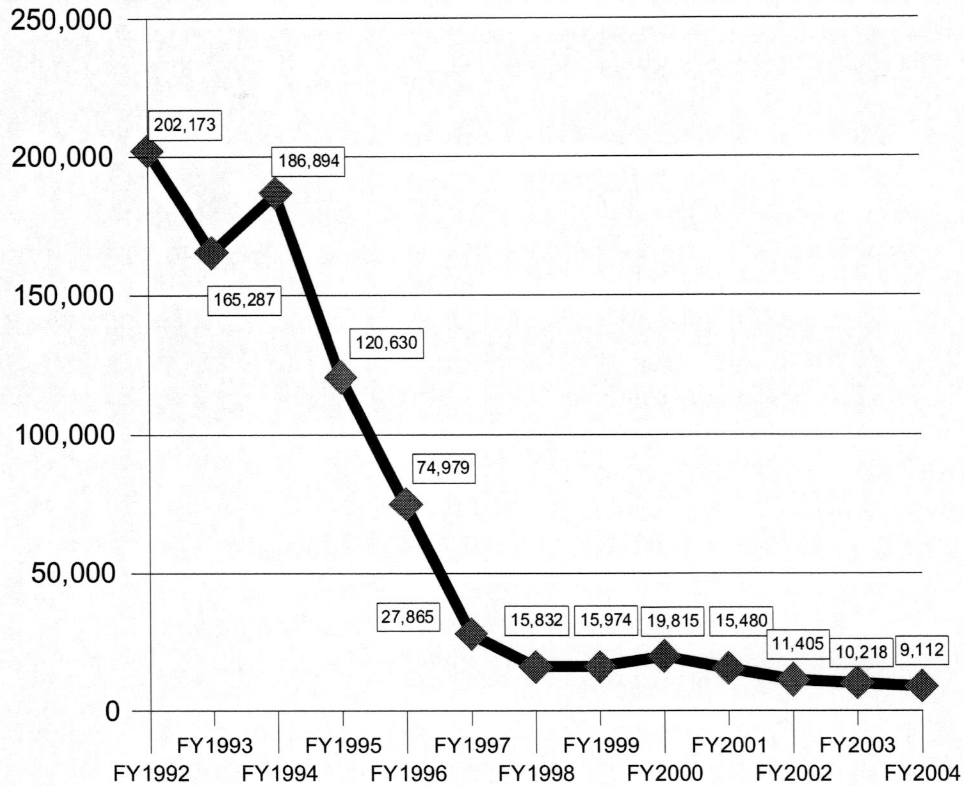

Source: CRS analysis of CBP data.

Figure 2 shows the apprehensions at the Chula Vista station over the same period of time. The trend in apprehensions at Chula Vista is somewhat similar to Imperial Beach, with overall apprehensions dropping significantly *from* FY1992 *to* FY2002. Apprehensions increased slightly from FY2002 to FY2004, but remain far below their early 1990s levels. Interestingly, the rate of decline in Chula Vista in the mid-1990s lagged behind the rate of decline in Imperial Beach station during this period. This suggests that as enforcement ramped up in Imperial Beach station, unauthorized migration shifted westward to Chula Vista. From FY1992 to FY1998, for example, apprehensions decreased by 92% in Imperial Beach, but only by 54% in Chula Vista. From FY1998 through FY2001, apprehensions leveled off in Imperial Beach, averaging around 16,000 a year, but continued to decline at Chula Vista, from 72,648 in FY1998 to 3,080 in FY2002. Overall, the trend indicates the following: as enforcement measures, in this case including fencing, were deployed — first focusing on Imperial Beach, and later extending to Chula Vista — the flow of unauthorized migration pushed eastward. The drop in apprehensions occurred first in Imperial Beach, and then later pushed eastward to Chula Vista.

Figure 2. Chula Vista Station Apprehensions

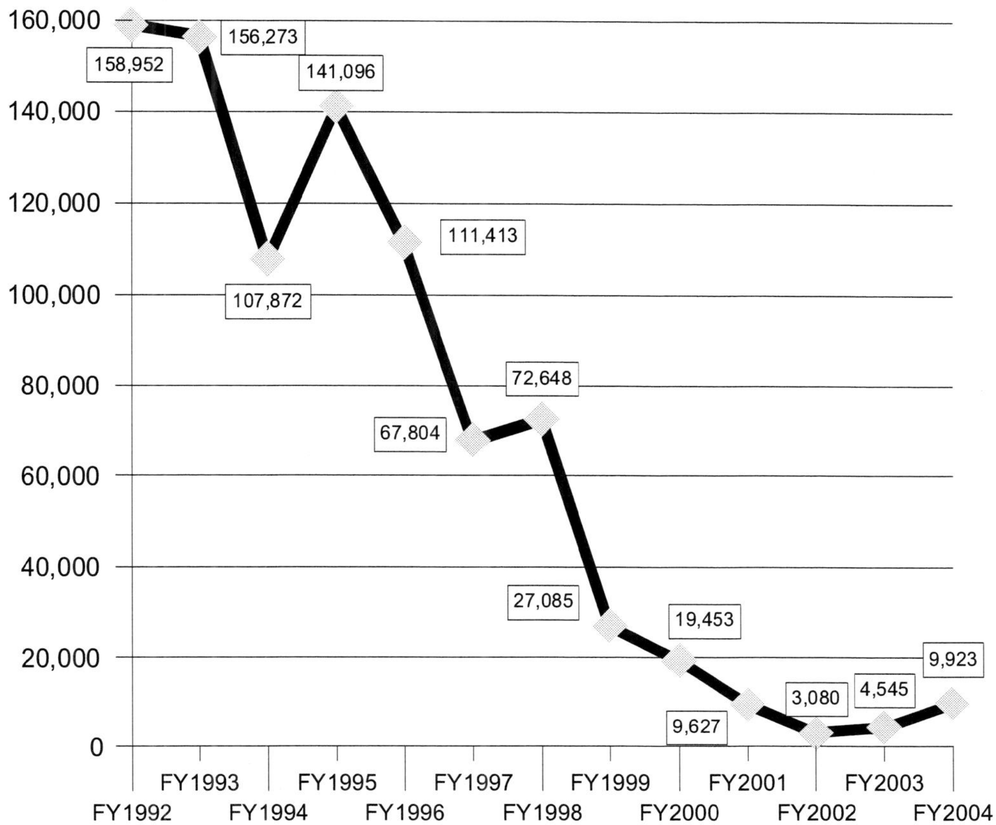

Source: CRS analysis of CBP data.

Figure 3 shows the aggregate apprehensions made at the other San Diego sector stations, excluding Imperial Beach and Chula Vista. Those stations are El Cajon, Campo, San Clemente, Temecula, and Brown Field. **Figure 3** shows that at the time apprehensions were beginning to decline in Imperial Beach (starting in FY1995) and Chula Vista (starting in FY1996), apprehensions at other San Diego sector stations almost doubled. This suggests that as enforcement efforts increased in the two westernmost stations, including the installation of fencing and the deployment of additional agents, the flow of illegal migration pushed eastward to the other stations in the San Diego sector. While apprehensions declined in the non-fenced stations of the San Diego sector from FY1997 to FY2001, the rate of decline was not as steep as the rate of decline at the stations where fencing was deployed. Overall, the decline in apprehensions in the rest of the San Diego sector has lagged behind the decreases in Imperial Beach and Chula Vista: from FY1992 to FY2004, apprehensions in the other San Diego sector stations decreased by 42%, compared to decreases of 95% in Imperial Beach and 94% in Chula Vista. In FY2003 and FY2004, apprehensions increased slightly in the rest of San Diego sector, possibly in response to the increasing USBP focus on the Tucson sector in Arizona.[47] It seems, then, that the

[47] For more information on overall apprehension trends, please refer to CRS Report RL32562, *Border Security: The Role of the U.S. Border Patrol*, by Blas Nuñez-Neto.

installation of border fencing, in combination with an increase in agent manpower and technological assets, has had a significant effect on the apprehensions made in the San Diego sector. This in turn suggests that fewer unauthorized aliens are attempting to cross the border in the San Diego sector as a result of the increased enforcement measures, including fencing, manpower, and other resources, that were deployed to that sector.

Figure 3. Apprehensions at San Diego Sector Stations, Excluding Imperial Beach and Chula Vista

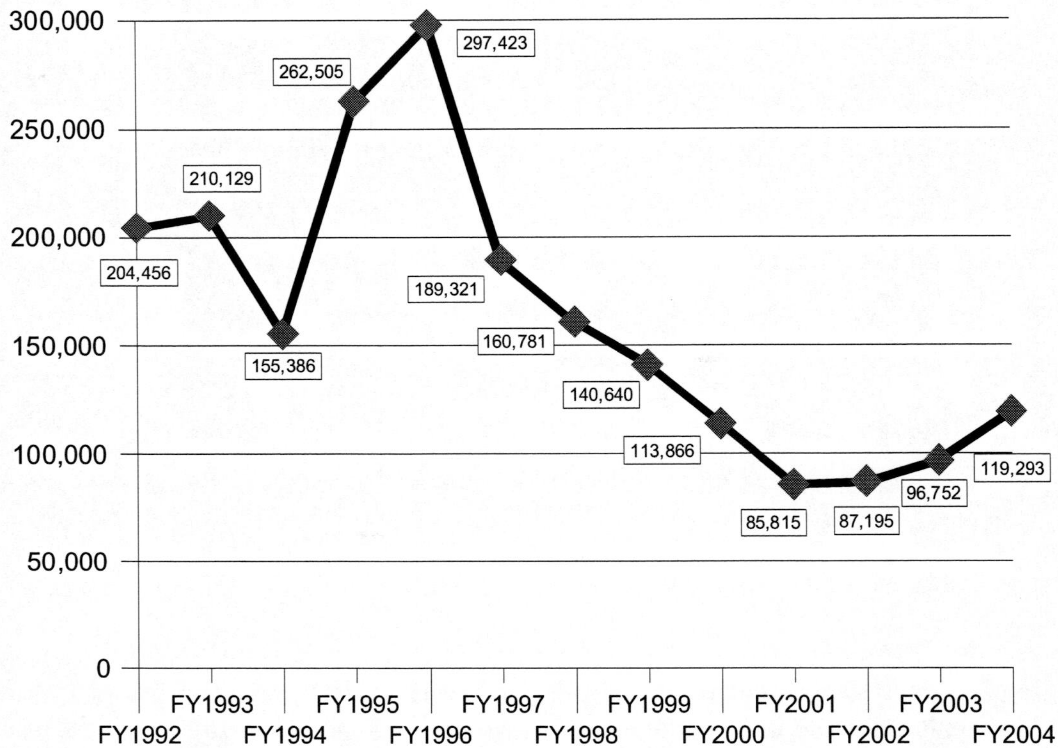

Source: CRS analysis of CBP data.

Figure 4 shows overall San Diego sector apprehensions, breaking out the Imperial Beach and Chula Vista stations, and compares them to the apprehensions made at the Tucson sector between FY1992 and FY2004. The data used to create this graph can be seen presented in table form in **Appendix V**. **Figure 4** shows that in FY1992, Imperial Beach and Chula Vista accounted for 64% of all apprehensions made in the San Diego sector; by FY2004 the two stations accounted for only 14% of all apprehensions made in the sector. However, as apprehensions declined in Imperial Beach and Chula Vista stations and San Diego sector as a whole over the late 1990s and early 2000s, apprehensions in the Tucson sector in Arizona increased significantly over this period. Over the 12-year period between 1992 and 2004, overall apprehensions in the San Diego sector declined by 76%. However, as apprehensions were decreasing in the San Diego sector, they were increasing in other sectors further east. This increase was most notable within the Tucson sector in Arizona, where apprehensions increased six-fold (591%) between FY1992 and FY2004. As **Figure 4** shows, overall apprehensions in the San Diego and Tucson sectors combined have averaged roughly 620,000 yearly since FY1992, with the San Diego sector accounting for the lion's share during the early 1990s and the Tucson sector accounting for the majority in the early 2000s. This provides further indication that the construction of the fence, combined with the increases in manpower in the San Diego sector, changed the patterns of migration for unauthorized aliens attempting to enter the country illegally from Mexico.

Figure 4. Apprehensions at San Diego Sector Stations and Tucson Sector

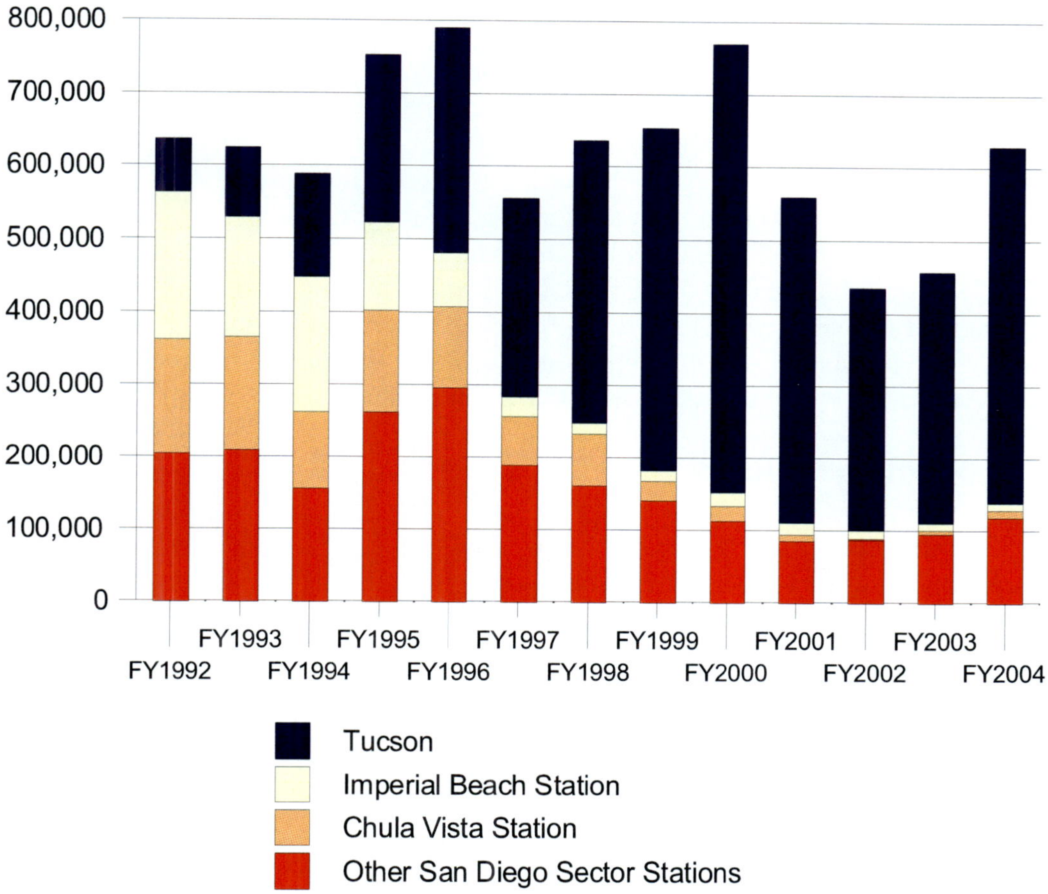

Source: CRS analysis of CBP data.

As **Figures 1-4** show, the increased deployment of agents, infrastructure, technology, and other resources within the San Diego sector has resulted in a significant decline in the number of apprehensions made in that sector. Nationally, apprehensions made by the USBP grew steadily through the late 1990s, only to decline in the early 2000s. However, in 1992 the USBP apprehended 1.2 million unauthorized aliens; in 2004, the USBP also apprehended 1.2 million unauthorized aliens.[48] While the increased enforcement in the San Diego sector has resulted in a shift in migration patterns for unauthorized aliens, it does not appear to have decreased the overall number of apprehensions made each year by USBP agents. As previously noted, apprehensions statistics can be somewhat misleading, but they

[48] CRS analysis of CBP data.

nevertheless remain the best way to codify trends in unauthorized migration along the border. However, it is impossible to ascertain solely by looking at apprehensions statistics how many unauthorized aliens are attempting to enter the country illegally, because it is unclear how many individuals evade being captured by the USBP each year.

Border Barrier Construction

The USBP has been constructing and maintaining barriers along the international land border since 1991. These barriers have historically been limited to selected urban areas as part of the USBP's overall strategy of rerouting illegal migration away from urban areas towards geographically isolated areas where their agents have a tactical advantage over border crossers. Two main types of border fencing have been constructed: primary fencing located directly on the border along several urban areas; and Sandia fencing, also known as secondary or triple fencing, in San Diego. Additionally, the USBP has begun installing permanent vehicle barriers in various segments of the border. Vehicle barriers are designed to impede the entry of vehicles while allowing individuals and animals to cross the border freely. As such, they have a lower environmental footprint than border fencing.

Steps Prior to Construction

Several considerations come into play whenever the USBP contemplates construction along the border. There are a number of steps that must be taken before the construction process can begin. These steps include, but are not limited to, determining what the environmental impact of the construction will be; acquiring the land needed for the fence; acquiring the materials that will be used for the fence; and securing the assistance of the Corps of Engineers and the National Guard for the construction process. The role the Corps of Engineers plays in assisting the USBP with th entire process of constructing border fencing, including acquiring materials, will be discussed subsequently in the construction process section. This section will cover the issues associated with environmental assessments and land acquisition.

Environmental Impact Assessments. Land along the southwest border supports a number of animals and plants and provides habitat to many protected species. The U.S. Fish and Wildlife Service, for example, reported that a total of 18 federally protected species have the potential to be found along certain sections of the California border.[49] In Arizona, at least 39 federally endangered, threatened, or candidate species can be found living along its border.[50] More than 85% of the lands directly along the Arizona border are federal lands, much of it set aside to protect wilderness and wildlife. For example, the Organ Pipe Cactus National Monument, the Cabeza Prieta National Wildlife Refuge, and the Buenos Aires National Wildlife

[49] EIS, San Diego Border Fence.

[50] Defenders of Wildlife, *On the Line — The Impacts of Immigration Policy on Wildlife and Habitat in the Arizona Borderlands*, 2006, p. 26. (Hereinafter, Defenders of Wildlife, *On the Line*.)

Refuge can all be found adjacent to the border. The southwest border region is considered a fragile environment, susceptible to harm from even the slightest changes to the ecosystem.[51]

Many are concerned with the geographic footprint and subsequent environmental impacts of both illegal immigration and USBP activities. Until the early 1990s, the USBP's enforcement activities along the border were nominal and the environmental consequences of illegal crossings went largely unnoticed. As illicit trafficking escalated, however, so did the USBP's activities and enforcement footprint, including the construction of fencing and other barriers. Although the San Diego fence reportedly reduced the number of aliens attempting to drive across the open border (and consequently the enforcement footprint to stop such activities), it did little to block the flow of foot traffic.[52] Illegal aliens often damage habitat by cutting vegetation for shelter and fire, causing wildfires, increasing erosion through repeated use of trails, and discarding trash.[53] Environmentalists claim that the USBP's enforcement activities, including the pursuit of illegal aliens, use of off-road vehicles and construction of roads and fences, compound the degradation.[54] The REAL ID Act will allow the DHS Secretary to waive any legal requirements needed to expedite the construction of border fencing. Until such time that DHS waives an applicable law, however, it must follow all legal requirements normally imposed on federal agencies, including, for example, NEPA documentary requirements.

Land Acquisition. The construction of a fence along the border necessarily requires the government to acquire some type of interest in the land. The San Diego border fence, for example, is to extend approximately 150-feet north of the international boundary.[55] Current immigration law authorizes the Secretary of DHS to contract for and buy any interest in land adjacent to or in the vicinity of the international land border when the Secretary deems the land essential to control and guard the border against any violation of immigration law.[56] It also authorizes the Secretary to accept any interest in land along the border as a gift and to commence condemnation proceedings if a reasonable purchase price can not be agreed upon. With respect to the San Diego border fence, the law requires the Secretary to

---

[51] Eilene Zimmerman, SFGate.com, *Border protections imperil environment — Last wilderness area south of San Diego could be damaged*, February 27, 2006, at [http://www.sfgate.com/cgi-bin/article.cgi?file=/c/a/2006/02/27/MNG2GHFBFL1.DTL &type=printable].

[52] EIS, San Diego Border Fence, at 1-10.

[53] Id. at 1-11.

[54] See generally, Defenders of Wildlife, *On the Line*, p. 26.

[55] Letter from Peter C. Sornsen, Acting Field Supervisor, U.S. Dept. of the Interior, to James Caffrey, Acting Director, Facilities & Engineering Division, Immigration and Naturalization Service, Re: Endangered Species Consultation for the Proposed 14-Mile Border Infrastructure System (July 1, 2003) (on file with author).

[56] 8 U.S.C. §1103(b).

promptly acquire such easements as necessary to implement the statute.[57] If DHS exercises its eminent domain powers, it must provide just compensation as required by the Constitution. In the case of the San Diego fence, construction of the final 4.5 miles continues to be held up as DHS acquires the necessary land.

DHS is authorized to acquire new interests in lands under the INA. However, the federal government may already own some land along the border pursuant to presidential proclamations made long ago. In 1907, President Roosevelt reserved from entry and set apart as a public reservation all public lands within 60-feet of the international boundary between the United States and Mexico within the State of California and the Territories of Arizona and New Mexico.[58] Known as the "Roosevelt Reservation," this land withdrawal was found "necessary for the public welfare ... as a protection against the smuggling of goods." The proclamation excepted from the reservation all lands, which, as of its date, were (1) embraced in any legal entry; (2) covered by any lawful filing, selection or rights of way duly recorded in the proper U.S. Land Office; (3) validly settled pursuant to law; or (4) within any withdrawal or reservation for any use or purpose inconsistent with its purposes. A similar reservation was made by President Taft in 1912, for all public lands laying within 60-feet of the boundary line between the United States and Canada.[59] This proclamation states that the customs and immigration laws of the United States could be better enforced and the public welfare thereby advanced by the retention in the federal government of complete control of the use and occupation of lands abutting the international boundary lines. The proclamation also provides exceptions similar to those described in the Roosevelt Reservation.

Border Fence Construction Process and Funding

CBP currently constructs border fencing under a Memorandum of Agreement (MOA) with the ECSO (Engineering and Construction Support Office) of the U.S. Army Corps of Engineers (Corps). ECSO manages several components of the construction process for CBP, including planning and acquisition of real estate; drafting the environmental protection plan; designing the project and formulating the engineering costs; overseeing the construction process; and enforcing the appropriate warranties. On most of the tactical infrastructure projects, National Guard units and military units from the Department of Defense (DOD) Joint Task Force North provide the labor. DOD uses these projects as part of their training regimen, leveraging their ability to deploy tactical infrastructure and thereby providing zero

---

[57] 8 U.S.C. §1101 note (b)(2).

[58] 35 Stat. 2136. The reservation also extends sixty-feet from the margin of any river that forms the international boundary. This language, however, does not apply to lands that abut the Rio Grande River in Texas since there are no federal "public lands" in Texas. Title to most of the western territories was obtained by the United States from foreign powers through purchase and treaty. Generally, the terms of acquisition provided for recognition of the few existing private property rights, but granted title over the vast non-private lands to the United States. Texas was an exception; it was admitted by annexation in 1845, and retained title to all its public lands. See United States v. Denver, 656 P.2d 1, 5 n.2 (Co. 1982).

[59] 37 Stat. 1741.

labor costs to CBP.[60] The funding for land acquisition and fence materials comes out of the CBP construction account within the DHS appropriation. Specific funding for fence construction is rarely identified in the conference reports, though it typically has been identified within the DHS (and previously the former INS) Congressional Budget Justifications.[61] **Table 1** shows the overall amount appropriated for the USBP construction account, and the specific amounts identified for tactical infrastructure within that account, since FY1996. Appropriations for fencing and other border barriers has increased markedly over the past five years, from $6 million in FY2002 to $647 million in FY2007. The FY2008 request, according to CBP, would include $196 million for fence construction.

Table 1. Border Patrol Tactical Infrastructure Appropriations

(millions of dollars)

Fiscal Year	Construction Account (total)	Tactical Infrastructure Construction
2008 (Request)	1,000[a]	196[a]
2007	1,500[b]	647[b]
2006	298	93
2005	92	15
2004	89	14
2003	235	23
2002	128	6
2001	133	3
2000	100	9
1999	90	4
1998	76	8
1997	10	4
1996	25	4

Sources: For FY2006, the amounts appropriated for construction and tactical infrastructure were identified from the FY2007 DHS Congressional Budget Justifications. For FY2004-FY2005, the amounts appropriated for construction and tactical infrastructure were identified from the FY2006 DHS Congressional Budget Justifications. FY2003 construction and tactical infrastructure funding was identified from the FY2005 DHS Congressional Budget Justifications. FY1996-FY2002 tactical infrastructure funding was identified in the FY2003 INS Congressional Budget Justifications; funding for FY1998-FY2000 includes San Diego fencing as well as fencing, light, and road projects in El

[60] Department of Homeland Security, Congressional Budget Justifications for Fiscal Year 2007, pg. CBP Construction 20. Hereafter referred to as DHS FY2007 Justifications.

[61] FY2006 is an exception. Within the conference report, $35 million was identified for the Southwest Border Fence and $35 million was identified for the construction of vehicle barriers and other border infrastructure in Tucson sector. H.Rept. 109-241.

Centro, Tucson, El Paso, and Marfa. FY2001 and FY2002 construction funding identified from the FY2002 INS Congressional Budget Justifications. FY2000 construction funding identified from the FY2001 INS Congressional Budget Justifications and H.Rept. 107-278. FY1999 construction funding identified from P.L. 105-277. FY1998 construction funding identified from P.L. 105-119. FY1997 funding identified from P.L. 104-208. FY1996 construction funding identified from P.L. 104-134.

Note: In FY2003 immigration inspections from the former INS, Customs inspections from the former customs service, and the USBP were merged to form the Bureau of Customs and Border Protection within DHS. As a result of this the data for years prior to FY2003 may not be comparable with the data for FY2004 and after. The USBP construction account has been used to fund a number of projects at the border, including fencing, vehicle barriers, roads, and USBP stations and checkpoints. In FY2007, the appropriations committee created a new Border Security Fencing, Infrastructure, and Technology (BSFIT) account within the CBP. This account funds the construction of fencing, other infrastructure such as roads and vehicle barriers, and border technologies such as cameras and sensors. Border fencing and infrastructure construction was transferred from the USBP Construction account to the new BSFIT account.

a. In FY2008, DHS requested $1 billion for BSFIT. According to CBP, $196 million of this request will be used for border fencing.[62]

b. The BSFIT appropriation in the FY2007 DHS Appropriation Act was $1.2 billion (see H.Rept. 109-699). Combined with the $300 million already appropriated in the emergency supplemental, the overall BSFIT appropriation for FY2007 was $1.5 billion. The appropriators did not offer guidance on how this funding was to be allocated between these different purposes. According to CBP, $647 million will be obligated for fencing in FY2007.[63]

Under the current MOA, once CBP purchases the materials and acquires the land, the Corps of Engineers undertakes the engineering studies and provides the manpower and machinery that are used to install the fencing. The actual manpower is typically provided by the State National Guard (the California National Guard, for example, constructed much of the San Diego fence), although occasionally the military, and sometimes the USBP, are involved in the construction.[64] The Corps of Engineers funding comes from the Department of Defense Drug Interdiction and Counter-Drug Activities Account. **Table 2** shows the funding for the "Southwest Border Fence" sub-account within this DOD Account, from FY1997 to FY2006.

---

[62] From e-mail correspondence with CBP, July 27, 2007.

[63] From e-mail correspondence with CBP, July 19, 2007.

[64] From interviews with CBP, November 30, 2005 and September 13, 2006, and the Corps of Engineers, November 29, 2005.

Table 2. DOD Funding for the Southwest Border Fence
(millions of dollars)

Fiscal Year	DOD Funding
2008	1.2
2007	N/A
2006	3.5
2005	N/A
2004	4.0
2003	4.7
2002	5.0
2001	5.0
2000	4.0
1999	3.0
1998	4.0
1997	5.0

Source: FY2008, H.Rept. 110-434; FY2007, H.Rept. 109-676; FY2006, H.Rept. 109-359; FY2005, H.Rept. 108-622; FY2004, H.Rept. 108-283; FY2004, H.Rept. 107-732; FY2002, H.Rept. 107-333; FY2002, H.Rept. 106-945; FY2000, H.Rept. 106-371, FY1999, H.Rept. 105-746; FY1998, H.Rept. 105-265; FY1997, H.Rept. 104-724.

Notes: N/A means not available. No funding was identified for border fencing in the FY2007 DOD Conference report, H.Rept. 109-676. The House Committee had recommended $8 million for this activity in H.Rept. 109-504, while the Senate Committee had not recommended any funding for it in S.Rept. 109-292. FY2005 funding for the "Southwest Border Fence" sub-account was also not identified in the DOD Conference Report, H.Rept. 108-622. The House Committee had recommended $7 million for this sub-account in H.Rept. 108-553; while the Senate Committee had not recommended any funding for it in S.Rept. 108-284.

Types of Fences and Barriers

The USBP currently uses three main types of barriers along the border: primary fencing immediately on the international border, Sandia fencing behind the primary fencing, and vehicle barriers meant to stop vehicles, but not people on foot, from traversing the border. While other forms of primary fencing, such as bollard fencing[65] and picket fencing,[66] have been constructed in limited areas,[67] historically

[65] Bollard fencing is comprised of vertical installations of solid concrete, metal spheres, or large posts, embedded into the ground at small enough intervals as to be impassable. Bollard fencing is difficult to compromise but expensive to install. See **Appendix I** for a depiction of bollard fencing.

[66] Picket fencing is comprised of metal stakes set sufficiently close together as to be impassable. See **Appendix I** for a depiction of picket fencing.

[67] Roughly 13 miles of these alternate forms of fencing have been constructed to date, according to an interview with CBP Congressional Affairs on September 13, 2006.

the agency has largely focused on using the landing mat fencing as a primary fence and the Sandia fence as a secondary fence.

Landing Mat Fencing. Landing mat fencing is composed of army surplus carbon steel landing mats which were used to create landing strips during the Vietnam War. The landing mats form panels 12 feet long, 20 inches wide, and 1/4 inch thick, which are welded to steel pipes buried 8 feet deep every 6 feet along the fence. Each mile of fencing requires the use of 3,080 panels.[68] There are about 5 miles of surplus landing mat fencing remaining as of 2006.[69] According to the USBP, sites that feature landing mat fencing include the following USBP stations: Campo, CA; Yuma, AZ; Nogales, AZ; Naco, AZ; Douglas, AZ, and El Paso, TX.[70]

In a 1999 study which was commissioned by the INS and performed under a Memorandum of Understanding, the Corps of Engineers predicted that construction costs for the landing mat fencing would range from $388,005 to $431,117 per mile.[71] This estimate includes the cost of materials, despite the fact that the landing mat fencing constructed to date has been comprised of army-surplus panels acquired by CBP at no cost. As previously noted, however, only about 5 miles of surplus landing mat fencing material remains available. Maintenance costs per year could vary widely depending on the number of breaches the fence undergoes. Low levels of damage to the fence would result in low annual repair costs, while a large number of breaches could result in stretches of fencing needing to be replaced. Per mile, the Corps of Engineers estimated that yearly maintenance costs would probably range from $1,742 to $17,753.[72] The Corps of Engineers noted that the net present value[73] of the fence after 25 years of operation would range from $5.4 million and $8.3

---

[68] U.S. Army Corps of Engineers, Construction Engineering Research Laboratories, *Engineering Life-Cycle Cost Comparison Study of Barrier Fencing Systems*, USACERL Technical Report 99/28, February 1999, p. 14. Hereafter referred to as Corps of Engineers Study.

[69] Interview with CBP Congressional Affairs, September 13, 2006.

[70] Telephone conversation with CBP, November 30, 2005.

[71] The Corps of Engineers used 1997 dollars in their study. For the purposes of this report, the numbers predicted by the Corps were adjusted to 2005 dollars using the Gross Domestic Product (GDP) deflator, available at [http://www1.jsc.nasa.gov/bu2/inflateGDP.html]. This website appears to be no longer operating; however, GDP deflator tables are also published by the Bureau of Economic Adjustment (BEA) at the Department of Commerce and are available at [http://bea.gov/bea/dn/nipaweb/TableView.asp?SelectedTable=13&FirstYear=1997&LastYear=2005&Freq=yr]. The actual predictions made by the Corps for constructing and maintaining primary fencing, in 1997 dollars, were $341,584 to $379,538 per mile for construction costs, and $1,534 to $15,629 per mile per year in maintenance costs. The 25-year life-cycle costs for constructing and maintaining landing mat fencing were predicted to range between $4,725,572 and $7,340,098 per mile in 1997 dollars.

[72] Corps of Engineers Study, p. 21.

[73] Net present value is a term used by the Corps of Engineers in their life cycle costs analyses for construction projects. It amortizes the future costs of a project and shows what the entire costs of the project will be. In this case, these numbers represent 25 year predictions and have been adjusted from 1997 dollars to 2005 dollars using a GDP Deflator

million per mile depending on the amount of damage sustained by the fencing each year.

Sandia Secondary Fence. The secondary fence proposed by the Sandia study has only been constructed over roughly 9.5 miles of the 14 miles in the original plan due to environmental concerns voiced by the California Coastal Commission. As previously discussed, P.L. 109-13 included language that will allow waiver of all legal requirements determined necessary by the Secretary of DHS for the expeditious construction of authorized barriers and only allows judicial review for constitutional claims. On September 14, 2005, DHS announced it is applying its new waiver authority to complete the San Diego fence.[74] However, construction has not begun on the remaining four miles of the San Diego fence because DHS is in the process of acquiring the necessary land.[75] DHS is currently estimating that it will cost an additional $66 million to finish the San Diego fence, bringing overall costs for this 14 mile-long project to $127 million. Additionally, DHS notes that it will use a mix of DOD resources and private contractors to finish the fence, and that the cost of using contractors is included in the request.[76]

The Sandia fence, as it has been constructed in the San Diego sector, is a secondary fence constructed behind the primary fence. Enough space is left between the two fences to accommodate an access road. The secondary fence is an angled two-piece fence. The fence is vertical up to ten feet high, and then extends out at an angle towards the climber. This prevents climbing by using gravity and the weight of the climber against them. The Corps of Engineers estimated that Sandia fencing costs per mile would range from $785,679 to $872,977 for construction and $953 to $7,628 per mile yearly for maintenance. Additionally, the Corps of Engineers study notes that the Sandia fence would possibly need to be replaced in the fifth year of operation and in every fourth year thereafter if man-made damage to the fence was "severe and ongoing." For this reason, in the study the Corps of Engineers noted that the net present value of the fence after 25 years of operation, per mile, would range from $11.1 million to $61.6 million.[77]

---

[74] DHS published a *Federal Register* notice on September 22, 2005, declaring the waiver of, in their entirety: (1) the National Environmental Protection Act (42 U.S.C. 4321 et seq.); (2) the Endangered Species Act (16 U.S.C. 1531 et seq.); (3) the Coastal Zone Management Act (16 U.S.C. 1451 et seq.); (4) the Federal Water Pollution Control Act (33 U.S.C. §§1251 et seq.); (5) the National Historic Preservation Act (16 U.S.C. §§470 et seq.); (6) the Migratory Bird Treaty Act (16 U.S.C. §§703 et seq.); (7) the Clean Air Act (42 U.S.C. §§7401 et seq.); and (8) the Administrative Procedure Act (5 U.S.C. §§551 et seq.).

[75] Telephone conversation with CBP, November 30, 2005.

[76] DHS FY2007 Justifications. pg. CBP Construction 18.

[77] The numbers used by the Corps of Engineers were cited in 1997 dollars. They have been adjusted to 2005 dollars using the GDP deflator cited above. The actual costs per mile in the Corps of Engineers Study were: $691,680 to $768,533 for construction, and $839 to $6,715 for maintenance. Net Present Value after 25 years in 1997 dollars ranged from $9.73 million to $54.23 million. Corps of Engineer Study, pp. 3 and 23.

Other Border Barriers: Vehicle Barriers

The USBP utilizes various different types of barriers to impede vehicles from crossing into the United States from Mexico. Some of these barriers are temporary and can be moved to different locations when needed, others are permanent barriers. The main purpose of vehicle barriers is to prevent smugglers from easily driving their vehicles across the border.

Permanent Vehicle Barriers. Permanent vehicle barriers, as their name suggests, are not designed to be moved but rather are permanent installations. Permanent vehicle barriers are typically steel posts, or bollards, that are excavated 5 feet deep and inserted into a poured concrete base. The posts alternate in above-ground height in order to dissuade individuals from forming a ramp over the barrier. They are spaced so as to allow foot and animal traffic but not vehicular traffic. The USBP recently began building permanent vehicle barriers in the Yuma sector, with a substantial stretch slated to be built along the Organ Pipe Cactus National Monument. When linked with the 30 miles of vehicle barriers built by the National Park Service, a USBP spokesman reportedly noted that the total 123 mile length of the project "will form the largest continuous physical barrier along the border in the nation."[78]

In the FY2007 DHS Congressional Budget Justifications, DHS notes that the Yuma vehicle barrier project would take until at least 2010 (and possibly longer) to complete if CBP continued to use the Corps of Engineers and other military personnel to construct the barriers. Instead, CBP proposes hiring commercial contractors to build 39 miles of vehicle barriers in the Yuma sector, or almost half of the project's 93 mile total.[79] CBP is projecting that the project will be completed by FY2011, and that the overall project costs will be $116 million.[80] This means that, overall, the project will cost roughly $1.25 million per mile. The National Park Service has spent $11.1 million to construct 18 miles of permanent vehicle barriers in Organ Pipe Cactus National Monument, and has obligated, but not yet spent, an additional $6.6 million in FY2005 funding to complete the remaining 13 miles of the project.[81]

---

[78] Jonathan Athens, "Officials say OK to Border Fence," YumaSun.com (July 20, 2005) available at [http://sun.yumasun.com/google/ysarchive14980.html].

[79] DHS FY2007 Justifications, pg. CBP Construction-7. CBP project length does not include the 30 miles of vehicle barriers maintained by the National Park Service.

[80] DHS FY2007 Justifications, pg. CBP Construction-18. It is unclear why the project is predicted to take less time with contractors, and yet the overall completion date for the construction is predicted to be 2011.

[81] From the National Park Service, February 9, 2006. The National Park Service notes that 30 miles of permanent vehicle barriers are being built at the Organ Pipe Cactus National Monument, and one mile is being built in the Coronado National Monument.

DHS currently has roughly 50 miles of vehicle barriers deployed along the border.[82] Vehicle barriers have been used in the El Centro, CA, Yuma, AZ, Tucson, AZ, and El Paso, TX sectors.[83]

Temporary Vehicle Barriers. Temporary vehicle barriers are typically built from welded metal, such as railroad track, but can also be constructed from telephone poles or pipe. These barriers are built so that they cannot be rolled or moved manually; they can only be moved with a forklift or a front-end loader. They are usually built at USBP stations and transported to areas of high vehicle entry, where they are placed and chained together.[84] The main advantage of the temporary vehicle barriers is their ability to be redeployed to different areas to address changes in smuggling patterns. The main disadvantage of these barriers is that they are easier to compromise than permanent vehicle barriers.

Current Status

In FY2007, DHS unveiled a new program, called SBInet,[85] that will deploy a mix of personnel, technology, infrastructure, and response assets in order to "provide maximum tactical advantage in each unique border environment." While SBInet has been billed as a nationwide initiative, its initial rollout has been confined to the southwest border. As part of SBInet, DHS awarded a contract to Boeing to serve as the project's lead technology integrator.

The SBInet program has included the construction of barriers as part of its approach to securing the border. Boeing, in conjunction with the Sandia National Laboratory, created a Fence Lab program to test the efficacy of 8 different fence designs.[86] In FY2007, CBP constructed a total of 76 miles of border fencing bringing the overall fencing at the border to 154 miles. In FY2008 through the end of the calendar year, CBP is planning to construct an additional 216 miles of fencing; this would bring the overall fencing at the border to 370 miles by the end of calendar year 2008.[87] The fencing that has been constructed thus far as part of SBInet has been primary fencing, and a few different designs have been used, including bollard fencing. While the National Guard was involved in some of the construction, much

[82] E-mail correspondence with CBP Congressional affairs, December 23, 2005.

[83] Telephone conversation with CBP, November 30, 2005.

[84] U.S. Department of Justice, Immigration and Naturalization Service, *Final Environmental Assessment U.S. Border Patrol Temporary Vehicle Barriers Naco and Douglas, Arizona*, November 2002.

[85] SBInet forms part of the Secure Border Initiative, which DHS has billed as a multifaceted approach to securing the border. DHS FY2007 Justifications, pg. CBP S&E-4.

[86] Department of Homeland Security, Customs and Border Protection, "Fence Lab To Test Effective Low Cost Solutions," *Secure Border Initiative Monthly*, April 2007.

[87] Presentation given by Rowdy Adams, SBI Deputy Executive Director, Department of Homeland Security, at the Border Management Summit, October 23, 2007.

of it was undertaken by contractors. In 2008, most of the fence construction will be done by contractors.[88]

In FY2007, CBP constructed 110 miles of vehicle barriers. CBP plans to build 190 additional miles of vehicle barriers by the end of calendar year 2008; this would bring the overall total milage of vehicle barriers at the border to 300.[89]

Legislation in the 110th Congress

The issue of border security continues to be of interest to the 110th Congress. The following sections describe notable legislation enacted or considered by Congress concerning barriers at the border.

Enacted Legislation

As part of the Consolidated Appropriations Act, 2008 (P.L. 110-161), enacted on December 26, 2007, Congress made a number of modifications to § 102 of IIRIRA, and significantly increased the Secretary of Homeland Security's discretion as to where to construct fencing along the southwest border. Whereas the Secretary was previously required to install roughly 850 miles of reinforced fencing along five stretches of the southwest border, a more general requirement has now been imposed on the Secretary to construct reinforced fencing:

> along not less than 700 miles of the southwest border where fencing would be most practical and effective and provide for the installation of additional physical barriers, roads, lighting, cameras, and sensors to gain operational control of the southwest border.[90]

The act further specifies that the Secretary of Homeland Security is not required to install

> fencing, physical barriers, roads, lighting, cameras, and sensors in a particular location along an international border of the United States, if the Secretary determines that the use or placement of such resources is not the most appropriate means to achieve and maintain operational control over the international border at such location.[91]

The act also amends the provisions of IIRIRA § 102 concerning fence construction in priority areas, by requiring the Secretary of Homeland Security to

---

[88] Ibid.

[89] Presentation given by Rowdy Adams, SBI Deputy Executive Director, Department of Homeland Security, at the Border Management Summit, October 23, 2007.

[90] P.L. 110-161, Div. E, § 564. Unlike under prior law, the Consolidated Appropriations Act, as enacted, does not specify that reinforced fencing be "at least 2 layers." See P.L. 104-208, Div. C, § 102(b), as amended by P.L. 109-367, § 3.

[91] Id.

identify either 370 miles or "other mileage" along the southwest border where fencing would be most practical and effective, and to complete construction of fencing in identified areas by December 31, 2008. This language replaces the prior language of IIRIRA § 102 concerning priority areas, which had been added by the Secure Fence Act and imposed deadlines for the construction of fencing and the installation of an interlocking surveillance camera system along specified border areas.

The Consolidated Appropriations Act does not modify the existing waiver provision or limitation on judicial review contained in IIRIRA § 102, but does impose new consultation requirements on the Secretary of Homeland Security when carrying out duties under this section, and conditions appropriations under the act upon compliance with these requirements. Specifically, the Secretary is required to consult with the Secretaries of the Interior and Agriculture, state and local governments, Indian tribes, and property owners "to minimize the impact on the environment, culture, commerce, and quality of life" in areas near where fencing is to be constructed. The act specifies that this consultation requirement does not create or negate any right to legal action by an affected person or entity.

The Consolidated Appropriations Act of 2008 provides a total of $1,225 million ($1.2 billion) for SBInet. This represents an increase of $225 million over the Administration's request, and the amounts recommended by the House and Senate-passed versions of H.R. 2638. Of the $1,225 million ($1.2 billion) provided by P.L. 110-161, $1,053 million is designated as emergency funding, and $172 million is comprised of regular appropriations. The $1,225 million ($1.2 billion) is apportioned as follows: $1,088 million ($1.1 billion) for development and deployment ($1,053 million ($1.1 billion) in emergency funding, and $35 million in regular appropriations); $73 million for operation and maintenance; and $64 million for program management. Funding for the construction of the border fence is included in the development and deployment activity in the BSFIT account. However, it is important to note that other items, such as cameras and sensors, are also funded under this activity. Currently available authoritative documentation does not provide funding details below the activity level. Therefore, the portion of this funding that would be specifically directed to the border fence cannot be precisely determined. P.L.110-161 also withholds $650 million of the funding provided for SBInet until an expenditure plan is received and approved by the Appropriations Committees.

Proposed Legislation

Prior to enactment of the Consolidated Appropriations Act, a number of other bills were introduced in the 110th Congress that would have expanded or underlined the Secretary of Homeland Security's authority to construct fencing at the border. H.R. 4192, H.R. 3638, and H.R. 2954 would direct the President to construct the fencing authorized by the Secure Fence Act. S. 2348 would authorize $3 billion in emergency funding for a variety of border security purposes, including the construction of 700 miles of fencing. S. 2294 and S. 1984 would call for the construction of 700 miles of fencing and 300 miles of vehicle barriers within two years of enactment. S. 1269 would call for the construction of double layer fencing along the border from the Pacific Ocean to the Gulf of Mexico. S. 330 would call for replacing existing fencing in Tucson and Yuma sectors with double layer fencing and constructing a total of 370 miles of fencing and 500 miles of vehicle barriers along the border.

The issue of barriers at the border has also been of interest to the 110th Congress as a component of the larger immigration debate. During May and June 2007, the Senate considered a number of comprehensive immigration reform measures (S. 1348, S.Amdt. 1150 to S. 1348, S. 1639), though cloture was unable to be achieved on any of these proposals. Both S.Amdt. 1150, as amended, and S. 1639, as introduced, included language concerning fencing at the border that was similar to that which was ultimately enacted as part of the Consolidated Appropriations Act for FY2008.[92]

Some comprehensive immigration reform proposals also provided that the construction of border barriers would serve as a trigger mechanism for broader immigration reform to occur. S.Amdt. 1150, as introduced, would have required the construction of 370 miles of fencing and 200 miles of vehicle barriers before some provisions relating to legalization, adjustment of status, and temporary workers could take effect. In addition, S.Amdt. 1150 would have amended § 102 of IIRIRA to expressly authorize the construction of the San Diego fence. During the initial Senate floor debate for S.Amdt. 1150, S.Amdt. 1172 was adopted by unanimous consent and amended the trigger mechanisms to require 300 miles of vehicle barriers. S. 1639, as introduced, included similar language to S.Amdt. 1150, as amended, concerning barriers at the border. S. 1639 would have required DHS to construct 370 miles of fencing and 300 miles of vehicle barriers as part of the trigger mechanisms required before some provisions relating to legalization, adjustment of status, and temporary workers could take effect. S. 1369 would also have expressly authorized the completion of the San Diego fence.

Legislation in the 109th Congress

The 109th Congress enacted three pieces of legislation concerning border fencing, and considered several more. The REAL ID Act (P.L. 109-13), as previously noted, expanded DHS' waiver authority to expedite the construction of border fencing. The Secure Fence Act of 2006 (P.L. 109-367) directed DHS to construct five stretches of border fencing totaling roughly 850 miles.[93] The FY2007 DHS Appropriations Act (P.L. 109-295) provided $1.2 billion for the installation of fencing, infrastructure, and technology along the border; $31 million of this total was designated for the completion of the San Diego fence.[94] In addition to these Acts, a number of bills with fencing related provisions were passed by the House and the Senate. H.R. 4437, which would have directed DHS to construct five stretches of fencing along the border, was passed by the House on December 16, 2005. S. 2611, which called for 370 miles of fencing to be constructed, was passed by the Senate on

---

[92] See S.Amdt. 1168 (adopted by unanimous consent and modifying S.Amdt. 1150 to S. 1348); S. 1639, § 103.

[93] From CBP Congressional Affairs, September 25, 2006.

[94] H.Rept. 109-699, p. 130.

May 25, 2006. S.Amdt. 4788 was added to the Department of Defense Appropriation bill, H.R. 5631, on August 2, 2006, and would have appropriated $1.8 billion to the National Guard for the construction of border fencing. H.R. 5631 was passed by the Senate on September 7, 2006; however, this fencing provision was stripped from the bill during conference.

P.L. 109-295, the FY2007 DHS Appropriations Act, provided $1.2 billion in funding for border fencing, infrastructure, and technology; combined with the supplemental appropriation provided by P.L. 109-234, the conferees noted that DHS would have $1.5 billion for border infrastructure construction in FY2007.[95] The conferees directed DHS to submit an expenditure plan for this funding within 60 days of the bill's enactment, and withheld $950 million of the funding until the plan is received and approved by the House and Senate Committees. However, the act did not place any restrictions on how DHS is to apportion this appropriation between fencing, infrastructure, and technology.

P.L. 109-367, the Secure Fence Act, originated in the House as H.R. 6061 and was passed on September 14, 2006. H.R. 6061 was passed by the Senate on September 29, 2006 and signed into law on October 26, 2006. The act directed DHS to construct two-layered reinforced fencing and additional physical barriers, roads, lighting, cameras, and sensors along five stretches of the southwest border. CBP has estimated that these stretches of fencing total roughly 850 miles[96] of the southern border. The five stretches of the border that DHS was required to fence were the 20 miles around Tecate, CA; from Calexico, CA to Douglas, AZ; from Columbus, NM to El Paso, TX; from Del Rio, TX to Eagle Pass, TX; and from Laredo, TX to Brownsville, TX. The act designated the roughly 370 mile portion of the fence between Calexico, CA, and Douglas, AZ, a priority area and directed DHS to ensure that "an interlocking surveillance camera system" is installed along this area by May 30, 2007, and that the fence is completed in this area by May 30, 2008. The act also designated a 30-mile stretch around Laredo, TX, as a priority area and directed DHS to complete this fencing by December 31, 2008. This language was similar to that passed earlier by the House in H.R. 4437. The fencing provisions in H.R. 4437 were largely identical to those in H.R. 6061, except that H.R. 4437 featured earlier construction deadlines for the priority areas identified by one year for the Calexico, CA, to Douglas, AZ, stretch of fencing and by two years for the 30-mile stretch around Laredo, TX.

In addition to the bills discussed above, there were a number of bills in the 109th Congress that would have expanded the current fencing and other forms of barriers at the international land border. Some of these bills would have required fencing to be constructed along the entire southwest border, others would have identified particular stretches of land which would receive fencing, and still others would have

---

[95] For more information about DHS Appropriations, please refer to CRS Report RL33428, *Homeland Security Department: FY2007 Appropriations*, Jennifer Lake and Blas Nuñez-Neto, Coordinators.

[96] From CBP Congressional Affairs, September 25, 2006.

called for studies to determine whether fencing is a cost-effective way of securing the border.[97]

Issues For Congress

Congress may consider a number of policy issues concerning the construction of barriers along the border, including, but not limited to, their effectiveness, overall costs compared with benefits, possible diplomatic ramifications, unintended consequences, and the locations in which they are to be constructed. Although these issues apply to all potential barriers at the border, due to the focus on border fencing in the current congressional debate, this section will focus its analysis on the potential policy issues surrounding the construction of fencing at the border.

Effectiveness

Proponents of border fences point to the substantial reduction in apprehensions along the San Diego sector as tangible proof that fences succeed in reducing cross-border smuggling and migration where they are constructed.[98] Opponents attribute part of the decrease in apprehensions to the increase in manpower and resources in the sector and, pointing to the increase in apprehensions in less-populated sectors, contend that the fence only succeeds in re-routing unauthorized migration and not in stopping it.[99] The USBP, for its part, states that border fencing is a force multiplier because it allows its agents to focus enforcement actions in other areas. The USBP has also stated that the fencing constructed in urban areas has helped reroute unauthorized migration to less populated areas where its agents have a tactical advantage over border crossers. As previously noted, the number of USBP apprehensions in 2004 were almost identical to the number of apprehensions in 1992; the main difference is that San Diego accounted for the majority of apprehensions in 1992, whereas in 2004 Tucson and Yuma sectors accounted for the majority of apprehensions.

A possible issue for Congress concerns the overall effectiveness of border fencing, especially if it is not constructed across the entire border in question. In the limited urban areas where border fencing has been constructed, it has typically reduced apprehensions. However, there is also strong indication that the fencing,

---

[97] Bills with border fencing language in the 109th Congress included H.R. 418, H.R. 1268, H.R. 4083, H.R. 4312, H.R. 4313, H.R. 4437, H.R. 5067, H.R. 5456, H.R. 5631, H.R. 6061, S. 1916, S. 2049, S. 2061, S. 2117, S. 2368, S. 2377, S. 2454, S.Amdt. 3192, S. 2611, S. 2612, S. 3564, and S.Amdt. 4788.

[98] For the views of supporters of border fencing, refer to "We Need a Fence," available at [http://www.weneedafence.com/], and Thomas Sowell, "Let's Get Our Terms Straight," available at [http://www.annistonstar.com/opinion/2006/as-insight-0402-0-6d01s3130.htm].

[99] For the views of opponents of border fencing, refer to Eilene Zimmerman, "Against the Wall," *Salon*, December 12, 2005, at [http://dir.salon.com/story/news/feature/2005/12/12/border_wall/index.html], and Molly Ivins, "Another Brick in the Wall," available at [http://www.annistonstar.com/opinion/2006/as-insight-0402-0-6d01s3130.htm].

combined with added enforcement, has re-routed illegal immigrants to other less fortified areas of the border. Additionally, in the limited areas where fencing has been erected, there have been numerous breaches of the border fencing and a number of tunnels discovered crossing underneath the fencing. It stands to reason that even if border fencing is constructed over a significant portion of the land border, the incidences of fence breaches and underground tunnels would increase. Possible policy options to address these issues could include mandating that border fencing be highly tamper-resistant or directing CBP to invest in tunnel-detection technologies.

Costs

Because border fencing is a relatively new and limited phenomenon along the U.S.-Mexico border, there is a dearth of information concerning its overall costs and benefits. The Corps of Engineers study predicted that the costs of constructing a double layer fence consisting of primary fencing and Sandia fencing would range from $1.2 million to $1.3 million a mile, excluding the costs of land acquisition. The Corps of Engineers also predicted that the 25-year life cycle cost of the fence would range from $16.4 million to $70 million per mile depending on the amount of damage sustained by the fencing.[100] If significant portions of the border were to be fenced, reducing the areas along which individuals could cross the border, it may stand to reason that the fencing will be subjected to more breaches and other attempts to compromise than the fencing that has already been constructed. This may mean that the costs of maintaining border fencing that is widely deployed in the future will be higher than they have been thus far for the limited deployment. The Corps estimates do not include the costs of acquiring the land or most labor costs, since construction would be done by DOD; these could well turn out to be significant expenses if private contractors are used to construct the fencing as per DHS' FY2007 Congressional Budget Justifications. The Congressional Budget Office (CBO) has estimated that border fencing would cost $3 million a mile to construct and that maintenance would total roughly 15% of the overall project costs per year.[101] However, the CBO does not elaborate on what is included in those estimates. DHS predicts that the San Diego fence will have a total cost of $127 million for its 14-mile

[100] As previously noted on pages 19-21, these numbers reflect Corps figures for the construction and 25-year life cycle costs associated with erecting primary landing mat and secondary Sandia fencing along the border. The Corps study used 1997 dollars, which have been adjusted by CRS using a GDP deflator to 2005 dollars. The actual predictions made by the Corps for constructing and maintaining primary fencing, in 1997 dollars, were $341,584 to $379,538 per mile for construction costs, and $1,534 to $15,629 per mile per year in maintenance costs. The 25-year life-cycle costs for constructing and maintaining landing mat fencing were predicted to range between $4.73 and $7.34 million per mile in 1997 dollars. The actual predictions made by the Corps for constructing and maintaining Sandia fencing, in 1997 dollars, were: $691,680 to $768,533 per mile for construction, and $839 to $6,715 per mile for maintenance. The 25-year life cycle costs for constructing and maintaining Sandia fencing were predicted to range between $9.73 million to $54.23 million per mile in 1997 dollars. Corps of Engineer Study, p. 3 and pp. 21-23.

[101] Congressional Budget Office, *Congressional Budget Office Cost Estimate: S. 2611 Comprehensive Immigration Reform Act of 2006, as passed by the Senate on May 25, 2006*, August 18, 2006, p. 14, at [http://www.cbo.gov/ftpdocs/75xx/doc7501/s2611spass.pdf].

length when it is completed — roughly $9 million a mile. Construction of the first 9.5 miles of fencing cost $31 million, or roughly $3 million a mile, while construction of the last 4.5 miles of fencing is projected to cost $96 million, or roughly $21 million a mile.[102] However these costs may be somewhat misleading due to the following factors: construction of the fence was delayed for an extended period of time; the remaining construction involves filling a relatively large gulch which may be more complex than the average stretch of border; and DHS is proposing to use private contractors to expedite the construction process which may increase the labor costs and thus may increase the overall project costs.

Some have argued that building fences on the border is too expensive and would consume funding that would be better spent on hiring additional agents or deploying additional technologies to the border.[103] Others maintain that the costs of fencing are negligible compared to the costs of illegal immigration, and that fencing has been proven effective at decreasing illegal immigration in those areas where it has been deployed.[104] The USBP has testified that "for border control, for border security, we need that appropriate mix. It's not about fences. It's not about Border Patrol agents. It's not about technology. It's about all of those things."[105] At issue for Congress is how best to allocate scarce border security resources while safeguarding homeland security. Does border fencing represent the best investment of border security funding, and what is the appropriate mix of border security resources? How much will maintaining border fencing cost in the future, and which agency will be responsible for this maintenance? Will using private contractors to expedite the construction of border fencing increase or decrease the costs?

Fence Design

Congress mandated the design of the border fence in San Diego in IIRIRA. Many different fence designs could be deployed to the border, and each have their relative strengths and weaknesses. Concrete panels, for example, are among the more cost-effective solutions but USBP agents cannot see through this type of fencing; the USBP testified about their preference for fencing that can be seen through, so as to identify the activity occurring on the Mexican side of the border and thus preserve their tactical advantage over potential border crossers, and to better

---

[102] From the DHS FY2006 and FY2007 Congressional Budget Justifications.

[103] See Jason Ackleson, "Fencing in Failure; Effective Border Control is Not Achieved by Building More Fences," *Immigration Policy in Focus*, Vol. 4, Issue 2, April 2005, available at [http://www.ailf.org/ipc/policy_reports_2005_fencinginfailure.asp].

[104] For a series of examples, see Parapundit, *Immigration Border Control Archives*, available at [http://www.parapundit.com/archives/cat_immigration_border_control.html].

[105] Testimony of Kevin Stevens, Senior Associate Chief of Customs and Border Protection, in U.S. Congress, House Homeland Security Committee, Economic Security, Infrastructure Protection and Cyber Security Subcommittee, and House Government Reform Committee, Criminal Justice, Drug Policy, and Human Resources Subcommittee, *Fencing the Border: Construction Options and Strategic Placement*, 109th Cong., 2nd sess, July 20, 2006. Hereafter referred to as: *Fencing the Border* hearing, July 20, 2006.

avoid potential rockings[106] or other violent incidents. Sandia fencing has been effective in San Diego and can be seen through, but is among the more expensive fencing options. Bollard fencing has been effective in its limited deployment and can also be seen through, but is also expensive to install and to maintain. Chain link fencing is relatively economical, but more easily compromised.[107] If fencing is to be constructed along the border, an issue concerns what kinds of fencing should be constructed in order to maximize its deterrent effect and its utility to the USBP while minimizing the costs associated with its construction and maintenance.

Fence Location

The USBP has testified that border fencing is most effective for its operational purposes when deployed along urban areas.[108] In these areas, individuals crossing the border have a short distance to cover before disappearing into neighborhoods; once they have entered neighborhoods it is much more difficult for USBP agents to identify and apprehend unauthorized aliens. Also, from populated areas it is relatively easy for unauthorized aliens to find transportation into the interior. For these reasons, all of the border fencing constructed by the USBP to date has been built in urban areas abutting the border, such as San Diego, Nogales, and El Paso. In rural areas, the USBP testified that it has a tactical advantage over border crossers because they must travel longer distances before reaching populated areas. According to CBP, fencing is manpower intensive because agents must continually check the fence for breaches and for illegal activity. This does not represent a problem in urban areas, because the USBP stations are typically located near the border in those areas. In some of the more rural areas of the border, where the nearest towns and USBP stations may be many miles away from the border, this would mean that agents would need to spend much of their working day commuting from the nearest USBP station to the fence location.[109] Additionally, because the border fencing constructed to date has been built along urban areas it has been relatively easy to house the individuals involved in its construction. If border fencing is extended into the more remote areas of the border, the costs of its construction may increase due to the need to bring the individuals and goods needed to build the fence to these areas for extended periods of time. Lastly, some areas of the border are prone to severe weather effects, such as flash flooding, that could compromise any permanent structures constructed there.

A very practical issue concerns what areas of the border should be fenced. Should fencing be restricted to urban or semi-urban areas in order to give the USBP

---

[106] Rockings refer to the phenomenon of individuals on the Mexican side of the border hurling stones and other items over the fence at USBP agents and vehicles. In the Yuma sector, for example, agents patrolling along the fence are deployed in armored vehicles known as "war-wagons" to protect themselves from rockings and other forms of assault, which are common in that area. Information obtained during a CRS site visit to Yuma sector in August 2005.

[107] *Fencing the Border* hearing, July 20, 2006.

[108] *Fencing the Border* hearing, July 20, 2006.

[109] Interview with CBP Congressional Affairs, September 13, 2006.

a tactical advantage over border crossers, or should fencing be constructed along any geographical area of the border that features large numbers of unauthorized immigration? In rural areas, should fencing be limited to areas of high illegal entry in order to impede individuals from crossing the border, or should fencing be constructed as a deterrent in any area, even those featuring low levels of illegal entry? Should fencing be deployed in sectors where the distance between the nearest USBP station and the fence requires agents to spend most of their day commuting? Should fencing be deployed to the northern border as well as the southwest border? Will building fencing along more remote or environmentally harsher areas of the border increase the construction costs?

Land Acquisition

There are a number of issues associated with the acquisition of the land that would be required for border fencing. Much of the land along the California and Arizona border is owned by the federal government; however most of the land along the Texas border is owned by private individuals. What will the costs of acquiring the land to construct border fencing be, and have these costs been factored into estimates of border fencing costs? Will eminent domain be used to confiscate land from individuals who do not wish to have fencing built on their lands?

The reservations made by Presidents Roosevelt and Taft may have kept substantial parcels of land within the federal domain, depending mostly on the amount of public lands at the time and valid existing claims. CRS was not able to determine how many valid claims and land patents exist, if any, or the number of private developments that may be encroaching on the reservations. Nonetheless, it appears that only those who qualify under an exception or were provided land by statute have valid fee title claims within the reserved strip. If lands were mistakenly granted, sold, or transferred to private parties, these conveyances could be void because, as a general rule, rights can not be acquired in lands actually embraced in a legally valid withdrawal.[110] Compensation under the Fifth Amendment for private landowners may not be owed if private claims are not legitimate. Because the proclamations do not cite any supporting authority, some question the President's implied or inherent constitutional powers to issue them.[111] Others may argue that

---

[110] Charles F. Wheatley, *Study of Withdrawals and Reservations of Public Domain Lands,* at Vol. III, at A-7 (1969); *see also* Steel v. Smelting Co., 106 U.S. 447, 453 (1882) (observing that the patent, like the deed of an individual, is inoperative if the government never owned the property, or had previously conveyed it, or had reserved it from sale); United States v. Fennell, 381 F. Supp. 2d 1300 (D. N.M. 2005). *Cf.* United States v. California, 332 U.S. 19, 39 (1947) (finding the federal government's paramount rights in the three-mile belt along the California coast were not lost by reason of the conduct of its agents or the acquiescence of such agents in California's claim of title).

[111] See United States v. Midwest Oil, 236 U.S. 459, 471 (1915) (upholding the President's authority to make land withdrawals on the basis of implied acquiescence in such withdrawals by Congress), *repealed by* 43 U.S.C. §1714. The President's constitutional inherent withdrawal power derived from three theories — residual Executive power, stewardship, and constitutional necessity. See Wheatley, *Study of Withdrawals,* at Vol I, at 134. In *Midwest Oil*, the Court noted that by 1910, the President had implemented at least (continued...)

they conflict with the exclusive mandate given Congress by the Property Clause of the Constitution to regulate and dispose of federal property.[112] An issue for Congress may include whether these proclamations are, in fact, valid, and if so what actions are appropriate to take in the instances where individuals own land within the reservation's boundaries. Assuming the proclamations are valid, the reservations may provide the first 60 feet of necessary space for fence construction in many areas. However, the two layer fencing constructed to date includes 150 feet of land between its layers. An issue for Congress may involve whether to confine border fencing to the 60-feet easement reserved by the proclamations, or whether to acquire the additional 90 feet of land that would be needed to construct Sandia-style fencing.

A corollary issue may involve the authority of DHS to construct border fencing along tribal lands. The Arizona desert along the Tohono O'odham reservation has become one of the most heavily trafficked border areas in the country, and the USBP has been restricted in its operations in the reservation due to tribal concerns.[113] The Tohono O'odham have reportedly vowed to fight the construction of fencing on tribe-owned land, citing environmental and cultural concerns.[114] Under current law, the Secretary of the Interior may grant rights-of-way over and across tribal land, provided the Secretary receives prior written consent of the tribe.[115] If the tribe does not consent, DHS may look to its new waiver authority to construct a fence across tribal lands. It is unclear, however, whether the expanded waiver that was given to the Secretary of DHS would allow (or was intended to allow) the Department to override the statutory authority given to another federal agency. Ultimately, federal government holds all Indian lands in trust, and Congress may take such lands for public purposes, as long as it provides just compensation as required by the Fifth Amendment.[116]

Diplomatic Ramifications

The governments of Mexico and Canada have both voiced concern about the United States constructing barriers along the international border. Mexican President Vicente Fox has come out strongly against the construction of border barriers on numerous occasions, stating his belief that these projects isolate the two nations, create frustration and misunderstandings, and do not solve the underlying problems that lead individuals to enter the United States illegally. Mexican Press Secretary

---

[111] (...continued)
252 executive orders making reservations for useful, though non-statutory purposes. Id. at 471.

[112] U.S. CONST. Art. IV, §3, cl.2.

[113] The USBP has been prohibited from building permanent camera installations and from paving access roads leading to and along the border. Information obtained during a CRS site-visit to the Tohono O'odham reservation, August 2005.

[114] Randal Archibald, "Border Fence Must Skirt Objections From Arizona Tribe," *New York Times*, September 20, 2006.

[115] 25 U.S.C. §324.

[116] United States v. Sioux Nation of Indians, 448 U.S. 371 (1980).

Rubén Aguilar Valenzuela stated his government's belief that "history has also taught us that a wall is never the solution to problems and that all walls eventually get torn down."[117] The Mexican government has reportedly forwarded numerous diplomatic notes to the White House registering its complaints against the possible expansion of border fencing. The Canadian government has also reportedly voiced concern over legislative proposals that would require a study of fencing options along the northern border, citing the difficulties of fencing the northen border and the fact that the U.S. government has never discussed such a plan with Canadian authorities.[118] Deputy Assistant Secretary for Immigration and Customs Enforcement John P. Clark reportedly stated during Congressional testimony that the proposed expansion of border fencing "harkens back to the Chinese wall and the Berlin Wall, not the message we want to send to the Mexican government, the Canadian government, and the rest of the world."[119] There are a number of possible issues for Congress to consider involving the potential diplomatic ramifications of constructing barriers along the border: Do the gains in border security outweigh the risk of alienating Mexico and Canada? Should the Mexican or Canadian government's opinions or wishes be taken into account when border fencing is concerned? Given the need to coordinate intelligence and law enforcement activities at the border, should maintaining cordial working relationships with Mexico and Canada take precedence over sealing the border with physical barriers?

Environmental Considerations

A great deal of debate has been around the environmental impacts of border fencing. The addition of fences along the southwest border, according to some, could harm sensitive environments, adversely affect critical habitat for protected species, and block migratory patterns for animals. Indeed, these concerns were among the many voiced by the CCC in its objection to the completion of the San Diego border fence. After immigration officials, the CCC, and the environmental community could not agree on a fence design, Congress passed waiver language in the REAL ID Act that allows the Secretary of DHS to waive all "legal requirements" necessary to ensure expeditious construction of the barriers and roads in the vicinity of the U.S. border. The Secretary used this provision to waive a number of primarily environmental laws (see **Appendix I**) in order to complete the San Diego border fence. DHS maintains, however, that it will follow "best management practices" throughout construction and will be "mindful of the environmental impacts" that

---

[117] Mexican Government Press Release, "Crecimiento con Calidad: El Presidente Vicente Fox encabezará la cena de gala de la XI Cumbre Anual Hemispheria San Pedro 2005: Rubén Aguilar, Vocero de Presidencia," May 12, 2005. Translation by CRS. Available at [http://www.presidencia.gob.mx/actividades/crecimiento/?contenido=18195&pagina=31].

[118] Beth Gorham, "Canada Balks at U.S. Plan for Border Fence," *Canadian Press*, December 17, 2005, at [http://www.canada.com/nationalpost/story.html?id=6c13f3fd-bdfb-4346-99ef-3f01f870c801&k=60592&p=1].

[119] Eunice Moscoso, "Border Fence Would Cost Millions, Not Work Critics Say," *Cox News Service*, November 9, 2005.

might occur.[120] Nonetheless, the Secretary's broad waiver authority has many worried about potential fence projects along other areas of the southwest border. Some argue that a fence along the Arizona border could be especially destructive to endangered jaguar and Sonoran desert pronghorn populations that usually roam this area because it would fragment native habitat and ultimately reduce gene pools.[121] Officials from the U.S. Fish and Wildlife Service, however, have said that it is too early to speculate about the potential impact of a border fence on wildlife migration.[122] Others note that unauthorized migration negatively impacts the environment, and believe that the construction of fencing could actually have a beneficial impact for protected lands if it reduces the number of unauthorized migrants traversing through environmentally sensitive lands.

As Congress debates immigration reform and the addition of new border fences, Members will undoubtedly be called upon to balance national security interests with environmental protections. Because there does not appear to be a clear consensus on the environmental impacts of border fencing, there is some interest in a study of the issue.[123] The effects of the San Diego border fence, for example, may help scientists better understand and predict potential environmental consequences elsewhere. Should fencing be expanded along the southwest border, Congress may be interested in environmentally sensitive alternatives to normal fencing and whether they can effectively limit illegitimate cross-border traffic. Some argue that vehicle barriers may be less intrusive because they allow unimpeded wildlife movement but can limit damaging vehicular traffic.[124] Congress may also call on the Secretary to cooperate or coordinate certain activities with the environmental community, since the Secretary could waive many environmental requirements.[125]

---

[120] Eilene Zimmerman, SFGate.com, *Border protections imperil environment — Last wilderness area south of San Diego could be damaged*, February 27, 2006, available at [http://www.sfgate.com/cgi-bin/article.cgi?file=/c/a/2006/02/27/MNG2GHFBFL1.DTL &type=printable].

[121] Id; Defenders of Wildlife, *On the Line*, pp. 16-19.

[122] Chuck Mueller, Dailybulletin.com, *Experts say border fence would hurt bighorn sheep* (August 14, 2006) available at [http://www.dailybulletin.com/news/ci_4177153]. Reports also indicate that a constant flow of illegal aliens into the native habitat for these animals interferes with their use of certain lands and survival. See Defenders of Wildlife, *On the Line*, p. 18.

[123] Indeed, §129 of S. 2611, passed by the Senate in the 109th Congress, called on the Secretaries of the Interior, Agriculture, Defense, and Commerce, and the Administrator of the EPA to assess the environmental impacts, including the impact on zoning, global climate change, ozone depletion, biodiversity loss, and transboundary pollution, of physical barriers along the southern international land and maritime borders.

[124] Defenders of Wildlife, *On the Line*, p. 35; Anne Minard, National Geographic News, *U.S. Immigration Law Could Harm Desert Animals, Critics Say*, (March 31, 2006) available at [http://news.nationalgeographic.com/news/2006/03/0331_ 060331_desert_fence.html].

[125] See generally, Defenders of Wildlife, *On the Line.*

Legal Considerations

The building of barriers along the international border has raised a number of legal issues. Most stem from requirements posed by environmental laws. Before the passage of the REAL ID Act waiver provision, for example, the Sierra Club and other environmental groups challenged, under the National Environmental Policy Act, the federal government's plan to complete the San Diego border fence.[126] The lawsuit alleged, among other things, that the government's final environmental impact statement did not address the entire 14-mile border infrastructure system and inadequately addressed the parts that were evaluated. After Secretary Chertoff exercised the waiver authority, the court dismissed the environmentalists' lawsuit in December 2005.

With respect to the Secretary's use of the waiver authority, the provision allows legal redress only for constitutional violations and limits review to the district courts of the United States (though the Supreme Court retains discretionary appellate review over district court decisions). In essence, an individual could not sue DHS for bypassing the environmental impact statement requirements of the National Environmental Policy Act (a law it has waived) because that would be a statutory violation, but an individual could sue for the taking of property without "just compensation" as provided by the Fifth Amendment. Should a district court make a ruling, that decision can only be appealed to the Supreme Court — that is, there is no appellate court review. Appeal directly from a district court to the Supreme Court rarely appears in law,[127] and according to some scholars, has been a "failure."[128] Past experiences, for example, demonstrated that the cases took up a disproportionate amount of time for oral argument and came to the Court on inadequate records.[129] Still, when Congress determines a particular class of cases to be of great public import, it is not unprecedented for it to require prompt review in the highest court of the land. As previously discussed, the Sierra Club and Defenders of Wildlife brought suit in the U.S. District Court for the District of Columbia in late 2007, challenging the constitutionality of the waiver authority provided to the Secretary of Homeland Security by the REAL ID Act, but the court rejected plaintiffs' constitutional challenge and dismissed the claim.[130]

---

[126] *Sierra Club v. Ashcroft*, No. 04-CV-272, (S.D. Cal. February 10, 2004).

[127] Laws that allow a district court ruling to be appealed directly to the Supreme Court include 13 U.S.C. §141 (illegal use of census data); 15 U.S.C. §29 (Sherman Act violations); 18 U.S.C. §700 (flag desecration violations); 42 U.S.C. §1971 (voting rights violations); and 42 U.S.C. §2000a-5 (civil rights violations).

[128] Charles Alan Wright & Mary Kay Kane, Law of Federal Courts, Ch. 12, §105 (6th ed. 2002).

[129] Id. Moreover, 28 U.S.C. §1254 allows the Court to bypass the courts of appeals by granting certiorari before judgment in those courts.

[130] See *supra* at 10.

Unintended Consequences

Considerable evidence shows that the USBP's historical strategy of "Prevention through Deterrence," whereby agents and resources including border fencing and other barriers have been concentrated along urban areas and areas traditionally featuring high levels of illegal entry, has succeeded in changing the flow of illegal migration. While San Diego, CA, and El Paso, TX, were historically the two sectors that featured the most apprehensions and the highest levels of illegal immigration, since the mid-1990s and the advent of Operations Gatekeeper and Hold the Line in those sectors, the more remote geographical areas of the Arizona border have become the hot-spots for illegal migration into the United States. One unintended consequence of this enforcement posture and the shift in migration patterns has been an increase in the number of migrant deaths each year; on average 200 migrants died each year in the early 1990s, compared with 472 migrant deaths in 2005. Another unintended consequence of this enforcement posture may have been a relative increase, compared with the national average, in crime along the border in these more remote regions. While crime rates in San Diego and El Paso have declined over the past 15 years, the reduction in crime rates along the more rural areas of the border have lagged behind the national trends. Another unintended consequence of the border fencing has been the proliferation of tunnels dug underneath the border. In San Diego, where the double-layer Sandia fencing has been constructed, smugglers have dug numerous tunnels underneath the border fence. One such tunnel was almost a kilometer long and was built from reinforced concrete — evidence of a rather sophisticated smuggling operation.

A possible issue for Congress to consider as it debates expanding the existing border fencing is what the unintended consequences of this expansion could be. Given the re-routing of migration flows that have already occurred, are DHS and the relevant border communities prepared to handle the increased flow of illegal migration to non-reinforced areas? Is DHS prepared to deal with an increase in the phenomenon of cross-border tunnels and other attempts to defeat the purpose of the fencing? What will the impact on crime rates be along the unreinforced areas of the border? Will USBP agents be required to spend some of their patrolling time guarding the fence?

Appendix I: Examples of USBP Border Fencing

Bollard fence

Landing mat fence

Picket or decorative fence

Sandia fence

Source: U.S. Department of Justice, Immigration and Naturalization Service, *Environmental Assessment for Infrastructure Within U.S. Border Patrol Naco-Douglas Corridor Cochise County, Arizona*, August, 2000, p. 1-13.

Appendix II: The San Diego Fence

Source: U.S. Department of Homeland Security; *Environmental Impact Statement for the Completion of the 14-Mile Border Infrastructure System San Diego, California*, July 2003.

Appendix III: Permanent Vehicle Barrier Schematic

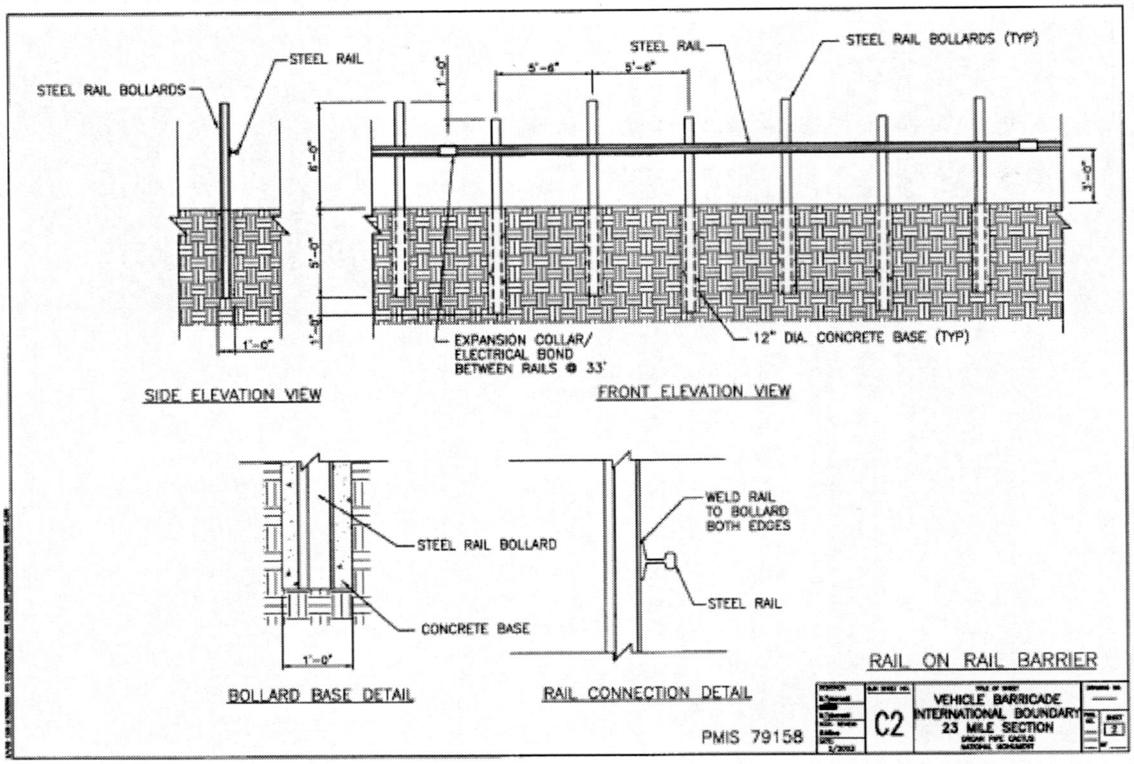

Source: U.S. Department of the Interior, National Park Service, *Proposed Vehicle Barrier Environmental Assessment*, April, 2003.

Appendix IV: Permanent Vehicle Barriers

Source: CBP Congressional Affairs.

Appendix V: Data From Figure 4

	FY1992	FY1993	FY1994	FY1995	FY1996	FY1997	FY1998	FY1999	FY2000	FY2001	FY2002	FY2003	FY2004
Other San Diego Sector Stations	204,456	210,129	155,386	262,505	297,423	189,321	160,781	140,640	113,866	85,815	87,195	96,752	119,293
Chula Vista Station	158,952	156,273	107,872	141,096	111,413	67,804	72,648	27,085	19,453	9,627	3,080	4,545	9,923
Imperial Beach Station	202,173	165,287	186,894	120,630	74,979	27,865	15,832	15,974	19,815	15,480	11,405	10,218	9,112
Tucson	71,036	92,639	139,473	227,529	305,348	272,397	387,406	470,449	616,346	449,675	333,648	347,263	490,827

Source: CRS Presentation of CBP data.

Appendix VI: Legal Requirements Waived by DHS for the Construction of the San Diego Border Fence

Laws Waived	General Requirements
National Environmental Policy Act (NEPA) 16 U.S.C. §§ 4321 *et seq.*	Under NEPA, an environmental impact statement must be prepared for "every recommendation or report on proposals for legislation and other major federal actions significantly affecting the quality of the human environment." If an agency is uncertain whether an action's impacts on the environment will be significant, it usually prepares an environmental assessment (EA). An EA is carried out to clarify issues and determine the extent of an action's environmental effects. NEPA also has public notice and comment requirements.
Endangered Species Act (ESA) 16 U.S.C. §§ 1531 *et seq.*	Section 7 of the ESA mandates that each federal agency consult with the Fish and Wildlife Service (FWS) or National Marine Fishery Services (NMFS), depending on the listed species involved, to ensure that its actions are "not likely to jeopardize the continued existence of any endangered species or threatened species, or result in the destruction or adverse modification of" designated critical habitat. Once consulted, FWS or NMFS must, if listed endangered species might be affected, prepare a *biological opinion* to determine the actual impact of the proposed action. Mitigation measures could be required.
Costal Zone Management Act (CZMA) 16 U.S.C. §§ 1451 *et seq.*	The CZMA requires federal agency activity within or outside the coastal zone that affects any land or water use or natural resource of the coastal zone to be carried out in a manner that is consistent to the maximum extent practicable with the policies of an approved state management program. The federal agency must submit a consistency determination to the applicable state agency.
Federal Water Pollution Control Act (Clean Water Act) 33 U.S.C. §§ 1251 *et seq.*	Section 404 of the Clean Water Act establishes a program to regulate the discharge of dredged or fill material into waters of the United States, including wetlands. Section 404 requires a permit before dredged or fill material may be discharged into waters of the United States, unless the activity is exempt.

Laws Waived	General Requirements
National Historic Preservation Act (NHPA) 16 U.S.C. §§ 470 *et seq.*	In accordance with the NHPA and its implementing regulations, 36 CFR Part 800, sites determined to be eligible for inclusion in the National Register of Historic Places must be protected, either through avoidance or other mitigative action, from direct and indirect impacts. The NHPA also has procedural requirements, including public notice and comment.
Migratory Bird Treaty Act (MTBA) 16 U.S.C. §§ 703 *et seq.*	Section 2 of the MTBA sets out the types of prohibited conduct and states: "Unless and except as permitted by regulations ... it shall be unlawful at any time, by any means, or in any manner, to pursue, hunt, take, capture, kill, attempt to do these acts, [or] possess ... any migratory bird, [or] any part, nest, or eggs of any such bird...." Violations of the MTBA may result in civil or criminal penalties.
Clean Air Act (CAA) 42 U.S.C. §§ 7401 *et seq.*	The Clean Air Act requires the Environmental Protection Agency to establish minimum national standards for air quality, known as National Ambient Air Quality Standards (NAAQS), and assigns primary responsibility to the states to assure compliance with the standards. Areas not meeting the standards, referred to as "nonattainment areas," are required to implement specified air pollution control measures. Federal actions located in NAAQS nonattainment areas must comply with the federal general air conformity rule set forth by the CAA and codified in 40 CFR Part 51. The general conformity rule requires federal agencies to ensure that actions undertaken in nonattainment or maintenance areas are consistent with the applicable state plan. The states administer the CAA through a comprehensive permitting program.
Administrative Procedure Act (APA) 5 U.S.C. §§ 551 *et seq.*	The APA establishes the general procedures that an agency must follow when promulgating a legislative rule. An agency must publish a notice of proposed rulemaking in the *Federal Register*, afford interested persons an opportunity to participate in the proceeding through the submission of written comments or, at the discretion of the agency, by oral presentation, and when consideration of the matter is completed, incorporate in the rules adopted "a concise general statement of their basis and purpose." A final rule must be published in the *Federal Register* "not less than 30 days before its effective date."

Appendix VII: Legal Requirements Waived by DHS for the Construction of Physical Barriers and Roads in the Vicinity of the Barry M. Goldwater Range in Southwest Arizona

Laws Waived	General Requirements
National Environmental Policy Act (NEPA) 16 U.S.C. §§ 4321 *et seq.*	See **Appendix VI** for description of requirements.
Endangered Species Act (ESA) 16 U.S.C. §§ 1531 *et seq.*	See **Appendix VI** for description of requirements.
Federal Water Pollution Control Act (Clean Water Act) 33 U.S.C. §§ 1251 *et seq.*	See **Appendix VI** for description of requirements.
Wilderness Act, 16 U.S.C. §§ 1131 *et seq.*	The Wilderness Act established a National Wilderness Preservation System on federal lands "where the earth and its community of life are untrammeled by man, where man himself is a visitor who does not remain." Within designated wilderness areas, section 4(c) of the act generally prohibits structures or installations, motor vehicle or other forms of mechanical transport, and temporary roads.
National Historic Preservation Act (NHPA) 16 U.S.C. §§ 470 *et seq.*	See **Appendix VI** for description of requirements.
National Wildlife Refuge System Administration Act, 16 U.S.C. §§ 668dd-668ee.	The National Wildlife Refuge System (NWRS) was primarily established to ensure the conservation of fish, wildlife, and plants. Designated areas may be used for other purposes (e.g., hunting, timber harvest, and grazing) only to the extent that such activities are compatible with the purposes for which the refuge was created. The refuges are managed by the Fish and Wildlife Service.

Laws Waived	General Requirements
Military Lands Withdrawal Act of 1999 (P.L. 106-65, 113 Stat. 885 (Oct. 5, 1999).	The Military Lands Withdrawal Act of 1999 withdrew the lands within the Barry M. Goldwater Range and generally reserved such lands to the Secretaries of the Air Force and the Navy for military purposes. The Secretaries of the Air Force, Navy, and Interior were required to establish an integrated natural resource plan (INRP) which, among other things, provided that "all gates, fences, and barriers constructed on such lands...be designed and erected to allow wildlife access, to the extent practicable and consistent with military security, safety, and sound wildlife management use."
Sikes Act, 16 U.S.C. §§ 670 *et seq.*	The Sikes Act requires the Secretary of Defense to carry out a program providing for the conservation and rehabilitation of natural resources on military installations (e.g., public lands withdrawn or reserved for use by a military department), pursuant to an INRP prepared in cooperation with the Secretary of the Interior.
Administrative Procedure Act (APA) 5 U.S.C. §§ 551 *et seq.*	See **Appendix VI** for description of requirements.

Appendix VIII: Legal Requirements Waived by DHS for the Construction of Physical Barriers and Roads in the Vicinity of the San Pedro Riparian National Conservation Area in Southeast Arizona

Laws Waived	General Requirements
National Environmental Policy Act (NEPA) 16 U.S.C. §§ 4321 *et seq.*	See **Appendix VI** for description of requirements.
Endangered Species Act (ESA) 16 U.S.C. §§ 1531 *et seq.*	See **Appendix VI** for description of requirements.
Federal Water Pollution Control Act (Clean Water Act) 33 U.S.C. §§ 1251 *et seq.*	See **Appendix VI** for description of requirements.
National Historic Preservation Act (NHPA) 16 U.S.C. §§ 470 *et seq.*	See **Appendix VI** for description of requirements.
Migratory Bird Treaty Act (MTBA) 16 U.S.C. §§ 703 *et seq.*	See **Appendix VI** for description of requirements.
Clean Air Act (CAA) 42 U.S.C. §§ 7401 *et seq.*	See **Appendix VI** for description of requirements.
Archeological Resources Protection Act (ARPA) 16 U.S.C. §§ 470aa *et seq.*	The Archeological Resources Protection Act generally prohibits the damage, removal, excavation, or alteration of any archeological resource located on public lands or Indian lands, except pursuant to a permit issued by the appropriate federal land manager.

Laws Waived	General Requirements
Safe Drinking Water Act (SDWA) 42 U.S.C. §§ 300f *et seq.*	The Safe Drinking Water Act provides federal authority for the establishment of standards and treatment requirements for public water supplies, control of the underground injection of wastes, and protection of sources of drinking water. Federal agencies involved in certain activities that may contaminate drinking water are subject to all federal, state, and local requirements concerning the protection of water systems to the same extent as any person is subject to such requirements.
Noise Control Act (NCA) 42 U.S.C. §§ 4901 *et seq.*	Pursuant to the Noise Control Act, the federal government has established standards for maximum sound levels generated from a variety of products. In addition, section 4 of the NCA requires federal agencies, subject to presidential exemption, to comply with federal, state, interstate, and local requirements respecting control and abatement of environmental noise to the same extent that any person is subject to such requirements.
Solid Waste Disposal Act (SWDA), as amended by the Resource Conservation and Recovery Act (RCRA) 42 U.S.C. §§ 6901 *et seq.*	Through the SWDA, as amended by RCRA, entities that transport or produce solid or hazardous waste are required to comply with regulations concerning the management, production, and storage of waste. Moreover, each federal agency engaged in any activity resulting, or which may result, in the disposal or management of solid waste or hazardous waste is subject to all federal, state, and local requirements concerning such waste to the same extent as any person is subject to such requirements.
Comprehensive Environmental Response, Compensation, and Liability Act (CERCLA) 42 U.S.C. §§ 9601 *et seq.*	CERCLA established broad federal authority to respond to the release or threatened release of hazardous substances. Among other things, it established requirements for closed and abandoned hazardous waste sites, and provided for liability of persons responsible for the release of hazardous waste at these locations. Federal agencies and instrumentalities are subject to these requirements to the same extent as nongovernmental entities, including with respect to liability.

Laws Waived	General Requirements
Federal Land Policy and Management Act (FLPMA) 43 U.S.C. §§ 1701 *et seq.*	The Federal Land Policy and Management Act establishes guidelines for the management and protection of federal public lands, as administered by the Secretary of the Interior through the Bureau of Land Management (in coordination with the Secretary of Agriculture with respect to lands in the National Forest System), and imposes procedural requirements for land transfers and exchanges. In developing land use plans, the Secretary is required to consider protection of areas of critical environmental concern and compliance with federal and state pollution control laws. The Secretary of the Interior, with respect to the public lands, and, the Secretary of Agriculture, with respect to lands within the National Forest System, are authorized to grant rights-of-way through such lands to other federal agencies, subject to terms and conditions imposed by the Secretary authorizing the right-of-way.
Fish and Wildlife Coordination Act (FWCA) 16 U.S.C. §§ 661 *et seq.*	The Fish and Wildlife Coordination Act generally provides that whenever the waters of any stream or other body of water are proposed to be modified by a federal agency, the agency must first consult with the United States Fish and Wildlife Service, Department of the Interior, and the head of the agency exercising administration over the wildlife resources of the state where the construction will occur, with a view to the conservation of wildlife resources.
Archaeological and Historic Preservation Act (AHPA) 16 U.S.C. §§ 469 *et seq.*	The purpose of the Archeological and Historical Preservation Act is to provide for the preservation of historical and archeological data which might otherwise be irreparably lost or destroyed as the result of, among other things, any alteration of terrain caused by a federal construction project. If a federal agency becomes aware that its activities in connection with a construction project may cause irreparable loss or destruction of significant scientific, prehistorical, historical, or archeological data, the agency must notify the Secretary of the Interior. If the Secretary deems such data to be significant and in danger of being irrevocably lost or destroyed, he is authorized to take action to protect and recover it.

Laws Waived	General Requirements
Antiquities Act 16 U.S.C. §§ 431 *et seq.*	The Antiquities Act authorizes the President to declare as national monuments historic landmarks, historic and prehistoric structures, and other objects of historic or scientific interest. This land is then withdrawn from any other use. The Secretaries of the Interior, Agriculture, and the Army may issue permits to qualified scientific or educational institutions for the excavation of archaeological sites and gathering of objects of antiquity on lands under their respective jurisdictions. Penalties are provided for damaging resources protected under the act.
Historic Sites, Buildings, and Antiquities Act (HSBAA) 16 U.S.C. §§ 461 *et seq.*	The Historic Sites, Buildings, and Antiquities Act declares it the national policy to preserve histories, sites, buildings, and objects of national significance. The Secretary of the Interior, through the National Park Service, is charged with implementing the policy of the HSBAA, including through the acquisition, maintenance, administration of historic sites. Persons who violate any rules or regulations promulgated under the HSBAA may be subject to a fine.
Arizona-Idaho Conservation Act of 1988 16 U.S.C.§§ 460xx *et seq.*	The Arizona-Idaho Conservation Act established the San Pedro Riparian National Conservation Area, consisting of public lands surrounding the San Pedro River in Cochise County, Arizona. The Secretary of the Interior is responsible for managing the area in a manner that conserves and protects its wildlife and other resources. The Secretary may only permit uses of the conservation area that are determined to further the primary purposes for which the conservation area was established. Except in limited circumstances, motorized vehicles are permitted only on designated roads. Persons who violate the act or its implementing regulations are subject to a fine and/or imprisonment.
Wild and Scenic Rivers Act 16 U.S.C. §§ 1281 *et seq.*	The Wild and Scenic Rivers Act establishes a National Wild and Scenic Rivers System (System) protecting rivers and adjacent lands with important scenic, recreational, geologic, fish and wildlife, historic, cultural, or other similar values. Components of the System are to be administered in a manner that protects and enhances the free-flowing and undeveloped nature of areas covered by the act.

Laws Waived	General Requirements
Farmland Protection Policy Act (FPPA) 7 U.S.C. §§ 4201 *et seq.*	The Farmland Protection Policy Act requires the Department of Agriculture, in cooperation with other federal entities, to develop criteria for identifying the effects of federal programs on the conversion of farmland to nonagricultural uses. Federal agencies are thereafter required to use this criteria to identify farmland that is converted by federal programs and take into account the adverse effects of such programs on the preservation of farmland. Agencies must consider alternative actions, as appropriate, that could lessen such adverse effects.
Administrative Procedure Act (APA) 5 U.S.C. §§ 551 *et seq.*	See **Appendix VI** for description of requirements.

INDEX

D

E

F

G